University Auditing in the Digital Era

Security, Audit and Leadership Series
Series Editor: Dan Swanson, Dan Swanson and Associates, Ltd., Winnipeg, Manitoba, Canada.

The *Security, Audit and Leadership Series* publishes leading-edge books on critical subjects facing security and audit executives as well as business leaders. Key topics addressed include Leadership, Cybersecurity, Security Leadership, Privacy, Strategic Risk Management, Auditing IT, Audit Management and Leadership.

The Security Hippie
Barak Engel

Finding Your Granite: My Four Cornerstones of Personal Leadership
Douglas P. Pflug

Strong Security Governance through Integration and Automation: A Practical Guide to Building an Integrated GRC Framework for Your Organization
Priti Sikdar

Say What!? Communicate with Tact and Impact: What to say to get results at any point in an audit
Ann M. Butera

Auditing Information and Cyber Security Governance: A Controls-Based Approach
Robert E. Davis

The Security Leader's Communication Playbook: Bridging the Gap between Security and the Business
Jeffrey W. Brown

For more information about this series, please visit: https://www.routledge.com/Internal-Audit-and-IT-Audit/book-series/CRCINTAUDITA

University Auditing in the Digital Era

Challenges and Lessons for Higher Education Professionals and CAEs

Sezer Bozkus Kahyaoglu
Erman Coskun

CRC Press
Taylor & Francis Group
Boca Raton New York London

CRC Press is an imprint of the
Taylor & Francis Group, an **informa** business

First Edition published 2022
by CRC Press
6000 Broken Sound Parkway NW, Suite 300, Boca Raton, FL 33487-2742

and by CRC Press
4 Park Square, Milton Park, Abingdon, Oxon, OX14 4RN

CRC Press is an imprint of Taylor & Francis Group, LLC

ISBN: 978-0-367-55322-7 (hbk)
ISBN: 978-1-032-22987-4 (pbk)
ISBN: 978-1-003-09300-8 (ebk)

DOI: 10.1201/9781003093008

Typeset in Sabon
by SPi Technologies India Pvt Ltd (Straive)

Contents

Abbreviations

1GU	First-Generation University
2GU	Second-Generation University
3GU	Third-Generation University
AA	Academic Analytics
APIs	Application Programming Interfaces
CAE	Chief Audit Executives
CAS	Course Authoring Systems
CKM	Customer Knowledge Management
CRM	Customer Relationship Management
COSO	Committee of Sponsoring Organizations
EDM	Educational Data Mining
HE	Higher Education
HER	Higher Education and Research
IAU	International Association of Universities
IEG	Independent Evaluation Group at World Bank
JRC	Joint Research Center of EU
KBE	Knowledge-Based Economy
KCI	Key Control Indicators
KPI	Key Performance Indicators
KRI	Key Risk Indicators
LA	Learning Analytics
LMS	Learning Management Systems
MOOCs	Massive Open Online Courses
STEM	Science, Technology, Engineering, Mathematics
STEAM	Science, Technology, Engineering, Arts, Mathematics
OECD	Organization for Economic Co-operation and Development
OKM	Organizational Knowledge Management
TEA	Technology-Enabled Auditing
UNESCO	United Nations Educational, Scientific and Cultural Organization

Foreword by Ali Nail Kubali, PhD

This book is the work of two outstanding academicians with ample practice and experience in the implementation of auditing and control systems.

The book does not only provide a solid theoretical foundation in the auditing of educational institutions, but also reflects the authors' knowledge accumulated over many years of auditing and management practice in profit- and nonprofit-making organizations.

The contents address two specific needs in the management of higher educational institutions.

The need for an up-to-date book for managers responsible for auditing and control of educational entities. Hence, the book will be a remarkable desktop reference document for these managers.

The acceleration in digitalizing of the management tools for procurement, production, storage, and service activities in the current decade has increased the need for compatibly digitalized control and auditing systems. This book successfully provides the digital interface between management and its auditing and control.

It is my opinion that this book will be of valuable help to build a more efficient, open, and transparent environment for all stakeholders of educational institutions.

Ali Nail Kubali, PhD
Chair and CEO,
NCM Consultancy

Foreword by Prof. Dr. Lourens Erasmus

Today's business practices are characterized by accelerating growth in the application of technology and "big data". It is almost unthinkable, nowadays, for any organization to function successfully without relying on its underlying information technology infrastructure – this is especially relevant to the higher education industry. That auditing in the higher education sector is regarded as a unique environment, is evident from the existence of the more-than-50-year-old Association of College and University Auditors (ACUA), operating mostly in North America. Acknowledging this niche area, the primary tenet of this book is how digital transformation is reshaping how higher education sectors emerge, operate, and evolve, and the response of auditors to this ever-challenging and evolving digital audit universe. Even though higher education is one of the industries anecdotally advancing the fastest toward technology-driven processes, the reality of COVID-19 accelerated the momentum. This book is published at a time of disruption, where the Coronavirus pandemic compelled universities, globally, to not only digitally transform their pedagogical practices to accommodate distance learning, but also to adopt virtual business processes and cloud storage, further necessitated by the new normal of non-office-based employment. A university's digital transformation impacts its risk profile, with previously identified risks possibly becoming obsolete, new emerging risks having to be incorporated and to be accounted for, and a possible shift in the severity of existing risks. In the same way that organizations' and especially the higher education industry's business models have been transformed, as a result of the increased use of technology and the ever-growing generation of and reliance on big data, how internal audit is practiced today has been impacted. This affects the internal audit plan and execution thereof and demands a new skill set from internal auditors. Apart from information technology-focused education and training, internal auditors require functional knowledge of the information technology environment, such as networks, databases, operating systems, and web-application security (cybersecurity), to become valuable members of the digitally transformed university's internal audit function. The comprehensive academic and professional work experience of the authors concerning internal auditing, information technology audit, digitalization, and academia, is evident from the pertinent information contained in the book. This book provides CAEs, internal auditors, risk officers, and compliance officers with insight concerning the rationale for the digitally transforming university, its digital environment, and the skills set and resources required by its internal audit function to be the value-adding function envisaged. It presents the internal auditor with knowledge regarding auditing in a digital environment and how to adapt to the audit culture in the quest to provide reliable assurance to stakeholders.

Prof. Dr. Lourens Erasmus
Department of Financial Governance
College of Accounting Sciences
University of South Africa

Authors

Dr. Sezer Bozkus Kahyaoglu, CIA, CFSA, CRMA, CFE, CICP, CPA graduated from Bosporus University in 1993 with an honor degree BSc in Management. Dr. Bozkus Kahyaoglu has an MA degree in Money Banking and Finance from Sheffield University and a Certification in Retail Banking from Manchester Business School, both with a joint scholarship from the British Council and the Turkish Bankers Association. After finishing her doctoral studies, she earned a Ph.D. in Econometrics from Dokuz Eylul University in 2015. In the period 1993–2004, Dr. Bozkus Kahyaoglu worked in the banking sector in various positions at the head office. Dr. Bozkus Kahyaoglu worked at Turkish Derivatives Exchange (TurkDEX-VOB) for two years as the founding member and head of the Audit and Investigations Department. In addition, Dr. Bozkus Kahyaoglu has worked in KPMG Risk Consulting Services as Senior Manager between 2007 and 2012. Afterward, Dr. Bozkus Kahyaoglu joined Grant Thornton as the founding partner of advisory services and worked there for two years in Business Risk Services. Afterward, Dr. Bozkus Kahyaoglu worked in SMM Technology and Risk Consulting as a Partner responsible for ERP Risk Consulting. During this period, Dr. Bozkus Kahyaoglu was a lecturer at Istanbul Bilgi University in the Accounting and Auditing Program for eight semesters (2008–2014) and the Ankara University Internal Control and Internal Audit Program for 2015–2017 for two semesters. Dr. Bozkus Kahyaoglu has been Associate Professor at Izmir Bakircay University from 2018 to the present. She has certifications of CIA, CFSA, CRMA, CICP, CFE, and CPA as well. Her research interests mainly include Applied Econometrics, Time Series Analysis, Financial Markets and Instruments, Energy Markets, Corporate Governance, Risk Management, Fraud Accounting, Auditing, Coaching, Mentoring, and NLP. Dr. Bozkus Kahyaoglu has various refereed articles, books, and book chapters published via international publishing companies. In addition, Dr. Bozkus Kahyaoglu is an active member of both IIA and ACFE, and supporting the profession by being a CREA member (IIA Global), a CBOK Steering Committee member (IIA Global), an Advisory Board member (ACFE Global), a Management Board Member (ACFE-Turkey Chapter), and a member of EMCC Turkey. Dr. Bozkus Kahyaoglu is married and has a daughter.

Dr. Erman Coskun is a full professor of Management Information Systems and Operations Research. He has an academic career of over 20 years, having held positions in the United States, Turkey, Cyprus, and the Kingdom of Saudi Arabia. He has taught courses focusing on Digital Transformation, Business Intelligence, Business Analytics, Global Connections Through Technology, Management Information Systems, Complexity and Complex Systems, Disaster and Risk Management, Supply Chain and Logistics Management, and Quantitative Methods and Management Science at bachelor, graduate, post-graduate, and executive levels. He has supervised the master's and PhD theses of over 20 students. He has won multiple teaching and career awards.

Dr. Coskun's research is focused on Business Analytics, Business Intelligence, Complexity and Complex Systems, ICT, Enterprise Resource Planning, and Disaster and Crisis Management. His research has been published in journals including *Government Information Quarterly*, *Systems and Software*, *Information and Software Technology*, and *Journal of Homeland Security and Emergency Management*, and in top conferences in MIS area such as International Conference on Information Systems (ICIS), Americas Conference on Information Systems (AMCIS), and European Conference on Information Systems (ECIS).

Dr. Coskun has held a variety of managerial and administrative positions including serving as the Founding Chair of Sakarya University's Management Information Systems department and program, the head of Sakarya University's Quantitative Methods department, deputy chair of Social Sciences Institute, AACSB Accreditation Committee, Sakarya University Rank and Tenure Committee, University Ethics Committee, Curriculum Committee, Business School Faculty Executive Board, Director of Social Sciences Institute and Cahir of Management Information Systems Departments at Bakircay University. He also has done consulting work for the private sector and government institutions in the United States and Turkey, where he has undertaken multiple projects to establish information systems and decision support systems and optimize analytical systems. He also served as a consultant for a financial credit scoring institution and an ERP software development company.

Dr. Coskun earned his doctorate in Engineering Science and his Master's in Industrial and Management Engineering from Rensselaer Polytechnic Institute in NY, USA. He also holds an MBA in Management Science and Quantitative Methods from Pace University, NY, USA, as well as an MS and BBA from Istanbul University, İstanbul, Turkey.

Dr. Coskun is currently serving as an Information Systems program evaluator for ABET and he is a member of the Association of Information Systems. He is also a founding member and academic relations committee chair of the Turkish chapter of AIS.

Introduction

It is a fact that digital transformation in universities is more than digitizing university documents, entering data into different types of information systems, and getting regular reports. It also includes humans. The key to success consists of joint work between the teams of administrative and academic staff, university management, and internal auditors, reaching effective communication among all participants of the process, including students and external stakeholders. To our knowledge, there are only a few publications on the auditing of the digital university. Hence, this book is aimed to contribute to the relevant literature by providing accurate and practical information to improve the audit quality and standards in the higher education process in today's digitally transforming world.

The coherent structure of a digitally transforming university suggests strategic forming of the digital infrastructure, learning environment, and innovative management culture. These elements are necessary for the full accomplishment of the digital university framework. Increased globalization of higher education, market pressure to undergo digital transformation in the university classrooms, and fast-changing needs and expectations regarding the e-learning environment are forcing auditors to rethink the new conditions in the audit universe and their audit culture. Hence, the auditors have additional encouragement for technology-enabled auditing (TEA) to fulfill their roles and responsibilities for providing reasonable assurance to stakeholders at the digital university.

Most books on internal audit deal with stated leading practice concepts and the IIA standards based on the classical approach. The concepts and methodologies are normally not issues from a professional viewpoint. The challenge comes into play when dealing with management, trying to balance the independence and objectivity of the position while also being officially employed by a digitally transforming institution. It is observed that sometimes CAEs become complacent in the requirement of professional skepticism and objectivity because of the push back they receive from management. In addition, although IIA's professional practices framework and international standards improved the audit culture, a significant number of Audit Committee and management are still unaware of the IIA standards and guidelines that internal auditors attempt to obey by when performing their fieldwork. In this book, the practical realities and lessons learned will be used as a CAE to apply real-life concepts to the challenges of managing an internal audit department in the digital era, gaining Audit Committee support by creating awareness about the digital risks. In this way, we provide relevant information about the major issues related to digitally transforming institutions for CAEs and internal auditors to become valued members of the team.

The main research question of the book is "How does digital transformation reshape the manner in which the higher education sector emerge, work, and evolve and in relation to this change in the workplace, what would be the response of the auditors to this challenging digital and more risky audit universe in transforming higher education sector?". In this respect, the main purpose of this study is to rethink the audit culture in the digital era and reveal the key characteristics that are open to improvement so that digitally transforming universities can be audited according to the higher education standards with a value-added audit and digitally supported approach. Based on this approach, the audit culture is reassessed, considering the digital university conceptual framework and business model. There are two main points to consider for the digital university work environment. The first is traceability and the second is auditability. In this respect, policy recommendations are made for best practices to achieve value-added digital audits in transforming universities.

After the COVID-19 pandemic period, it is a fact that Higher education is no longer an opportunity only for the cream of society or for those that live in advanced countries. Most of the colleges and universities around the world are opening their doors via the Internet, and similarly, countries that have previously resisted the idea of HE and e-learning are giving in to global economic pressure. In other words, they have started to accept the importance of life-long education and open particularly business schools. HE sectors around the world are aiming to mass-produce internationally trained professionals to meet the demands of an increasingly global business environment based on network organizations and platforms. In this respect, market-driven, scaled-down universities that focus on practical digital business skills are gaining importance and enrollment.

Considering the major market players in the HE sectors, nonprofit public and private universities in the United States dominate. On the other hand, there are some recent players entering the fray in the HE sectors. As more and more countries proceed toward a market economy structure, the expansion of the HE sector is usually the major tenet of their reforming attempts. As a result, more universities are opening in various regions such as Asia, Eastern, and Western Europe. For instance, in Asia, universities are struggling to rebuild their financial foundation after the economic reforms in the area. In addition, there are some new entrances of both for-profit entities and distance learning initiatives within the HE sectors. These HE institutions are mostly based in the United States, with some major players emerging from Western Europe.

As businesses around the world request high-tech workers, community and technical colleges are no longer the only market players in the HE sector. Private firms are more than happy to step in and educate workers – for a profit. For-profit education organizations and distance learning are the new competitive favorites for students. For-profit education organizations have huge resources and hence, they can conduct their businesses internationally thanks to the Internet.

It is a fact that off-campus learning is among the waves shaping the global higher education landscape. With technical advances and fast digitalization, HE institutions can offer education to executives, working adults, and others who do not have the time, money, or inclination to commit to the four-year, on-campus degree track.

The relationship between government and higher education is changing in various regions of the world, leading to strengthening the autonomy of HE institutions. Based on the advantages arising from autonomy, HE institutions are adopting more business-oriented personnel policies and greater market orientation. The key issue here is that "how does the higher education sector pay for new distance learning initiatives, much less the standard brick-and-mortar operations?" Due to the fundraising difficulties, consolidation is expected to accelerate in both the traditional brick and mortar universities and the for-profit sector, particularly in the distance learning sector. In this context, most of the colleges choose to cut a variety of administrative, maintenance, and recruiting costs by merging. As a result, most mergers, especially in the nonprofit education sector, where they prefer the term "joint affiliation", are hardly competitors.

In an economy, individuals who lose their jobs at the bottom of the business cycle start a new search. This search is an important factor that will determine the speed of turning back from the bottom of the business cycle. Innovations and the new situation tend toward the growth in the business cycle with the adaptation of individuals to these new conditions in the economy. In this respect, each economic crisis causes well-educated individuals to demand learning new things as well. Therefore, universities are the only institutions that determine the pace of exit from economic cycles. Especially, this situation will be more determinant in today's world where digital and digitalization-based industrial production structures develop.

The massive increase in production has accelerated as a result of the transfer of knowledge through universities. Since the existing knowledge is constantly transformed in universities, it is reflected in practice in a way that increases production even further. Universities accelerate the growth in production processes since they internalize technology. Especially, the development of digital technologies and the use of open-source software programs accelerate the application of theoretical knowledge in universities. At this point, the implicit knowledge through the strong human capital in the universities contributes to the application processes. In this way, the velocity of "implicit knowledge" becomes a determining variable in the digital transformation of universities.

The organization of the book is explained as follows:

Chapter 1 starts with the definition of the university from the past, present, and future perspectives. The characteristics of the universities, which have changed over time and are considered in three different generations, are briefly presented. In this process, the needs and requirements of change brought about by digitalization are emphasized. In this context, the digitalization strategy in the HE sector, digital business model, and new approaches starting with Industry 4.0 are explained. With the changing role of the HE institutions in the global value creation process, evaluations are made on Education 4.0. This change and development process in the HE sector also changes the institutional structure and culture. From this point of view, the contribution of internal auditors to this process and the requirements for harmonization of audit culture are emphasized.

Chapter 1 gives information about the conceptual framework of the digital university. The e-business model of digital university is introduced, and the digital business workflow is presented accordingly. Chapter 1 is essential for improving digital participation and for understanding the needs and expectations regarding the digitally transforming higher education (HE) institutions to adapt to new and challenging social contexts. There are various approaches for digital transformation such as Macro-, Meso-, or Micro-Engagements, which are applicable for the digital transformation of a university. Chapter 1 explains the key challenges and issues based on KPIs (key performance indicators), KRIs (key risk indicators), and KCIs (key control indicators) that the chief audit executives (CAEs) and HE professionals should be aware of, and emphasizes the digital risks that can be encountered in the digital transformation process, respectively. In particular, the responses of the auditors to the digital transformation process are examined and an answer is sought for the question of whether the auditors are ready by considering their strategic roles in this digital transformation.

Chapter 2 explains the business model and business processes of universities with a holistic approach. In this context, the internal control environment of the HE sector is defined and basic definitions of business processes, risk indicators, and performance indicators that are critical in terms of internal audit are explained. Practical information is provided on the importance of these business processes for internal audit and how audit tests should be carried out. This chapter contains key practical information that will form the basis of internal audit field works in terms of defining the basic information and risk factors specific to the HE sectors.

Chapter 3 focuses on the positioning of internal audit departments under the supervision of the rapidly digitizing HE sector in a globally competitive environment. It is expected that field works and audit tests, which form the basis of the internal audit, will now be performed in the virtual environment and with remote auditing techniques, i.e., CA/CM, instead of being carried out in physical environments. The main reason for this is that the work and almost all transactions of the auditees and process owners mostly take place in the digital environment. Therefore, internal audits should also adapt to the change process of the digital environment and carry out reasonable assurance and advisory services under international standards.

Chapter 4 discusses key issues that have arisen as a result of internal audit's evolution to digital audit. The important point here is that internal audit is now turning into digital audit and in some ways becomes a "trusted data analyzer". As a result, there must be a change in the structure of internal audit as a department and process. Recommendations for best practices are presented, with an emphasis on the challenges, opportunities, and threats created by this digital change. Information on the harmonization of internal audit charter, core competencies of internal audit, and service provision standards is provided. Considering IIA reporting standards, practices related to the establishment of digital reporting infrastructure are explained.

Chapter 5 includes information about the work, workforce, and the workplace, in which the change occurred with the effect of digitalization. Here, it is emphasized that internal audit should plan its core competencies and roles accordingly with the rapid digitalization and shift to distance education structure in the HE sectors. In addition, the strategic, operational, and legal roles of internal audits are explained. With digitalization, it is stated what the issues, competencies, and service delivery techniques and tools should be considered by internal auditors.

In Chapter 6, the requirements of a digitalized HE sector and the measurement criteria to ensure its effectiveness are examined. Accordingly, the size of the internal audit department, the competence of internal audit, the behavior of the auditees, institutional structuring, audit quality assurance, internal audit leadership structure, and the importance of communication and coordination with the Audit Committee are emphasized. Internal audit departments must provide services in the digital transformation process by considering the level of corporate maturity. In this context, it may be more appropriate to take an advisory role to monitor and support the effectiveness of change management in the digital transformation process. In this process, it should be ensured that an internal audit closely monitors and informs the management's risk perception.

Chapter 7 focuses on the effects of the COVID-19 pandemic that has accelerated the digital transformation process and made it imperative. In this context, a permanent change and transformation in the HE sectors along with developments in education curriculum and educational technologies are explained. The effects of new training techniques to increase the social benefit and the formation of digital communication platforms in a competitive environment are explained. To ensure that education is the new normal for lifelong and in all areas of life-wide, the transformation of the content and the application of educational technologies are revealed. In such a virtual environment, the audit universe and the risk universe are now becoming all in digital. This means that auditing priorities are changing. It is important to take precautions against cyber risks, especially considering the facts of the virtual environment. However, it should be taken into account that innovations such as digitalization, artificial intelligence applications, and blockchain can have devastating effects. To manage this process effectively, it is stated that the ethics audit will gain importance among audit types. Due to rapid technological developments and lagging legal regulations, it is recommended that ethical codes of conduct should be internalized in the corporate culture and it should be frequently audited.

Chapter 1

Conceptual framework for the digitally transforming university

Dynamic model approach

1.1 AN OVERVIEW OF THE CONCEPTUAL FRAMEWORK FOR THE DIGITALLY TRANSFORMING UNIVERSITY

After the emergence of the higher education (HE) institutions, the higher education system and institutions have experienced various "reformist" periods, starting from ancient times to the present. Nybom (2007) defines these reformist periods at HE institutions as the "HE revolution". These reformist periods are examined in six major phases[1] in the literature. Although the HE institution has its roots in ancient times, William von Humbolt in Germany led the foundation of modern research universities with scientific and organizational autonomy centered on the production of scientific knowledge in the 19th century (Ben-David and Zloczower 1962). For this reason, the periods that Välimaa (2014) expresses as Humbolt and modern research university periods are essential processes that are the continuation of each other. In this period, the number of universities has increased to meet the demand for engineers and scientists, which have increased with the industrial revolution, and universities have started to integrate into the industrial economy (Bates 2010). Thus, as the number of university graduates increased, the demand for universities increased rapidly, and HE has started to turn into a mass system. By the end of the 20th century, especially in developed and developing nations, it has been accepted that HE plays a key role in the growth of national economies with the emerging of the "knowledge-based economy" (KBE) concept as a result of significantly increasing globalization and availability of information technologies, It has also become an important topic of discussion in the international economy (George 2006). OECD (2002, 2020) makes an institutional classification to measure scientific and technological activities of sectors and they define the HE sector as follows:

> ...All universities, colleges of technology, and other institutions of post-secondary education, whatever their source of finance or legal status. It also includes all research institutions, experimental stations and clinics

DOI: 10.1201/9781003093008-1

> operating under the direct control of or administered by or associated with higher education institutions. ... The HE sector includes all establishments whose primary activity is to provide post-secondary (tertiary level) education regardless of their legal status.

Wissema (2009) defines this recent period as the "third-generation university-3GU" which is shown in Figure 1.1. In this respect, the 3GU approach is the starting point of this book to explain major needs and expectations for digitally transforming universities in the below sections.

It is a fact that the 3GU is both unavoidable and preferable because the trends that are destroying the 2GU approach are obvious and are summarized in Figure 1.2.

Due to globalization, there is a need for employment, expectation, and demand models at the international level. In this way, new approaches are introduced that can meet the multidisciplinary learning needs of individuals who have the flexibility to adapt to the fast-changing business environment (Riddell 1996). On the other hand, based on technological advancements, the emergence of educational platforms over the internet and online education opportunities have increased the number of students and the diversity of learning. New education models have also created new data sources on issues such as determining learning needs and evaluating educational outcomes and brought innovations. Because of all these developments, changes and transformations have started in many areas of HE, especially in the business model, strategy, scope, content, and financing of HE.

In a KBE, data are seen as a raw material to be converted into products, processes, or services, and well-educated people with advanced knowledge are needed to handle, analyze, and judge these data (Slaughter and Rhoades 2004). This situation is shaped by neoliberalist[2] policies accompanying

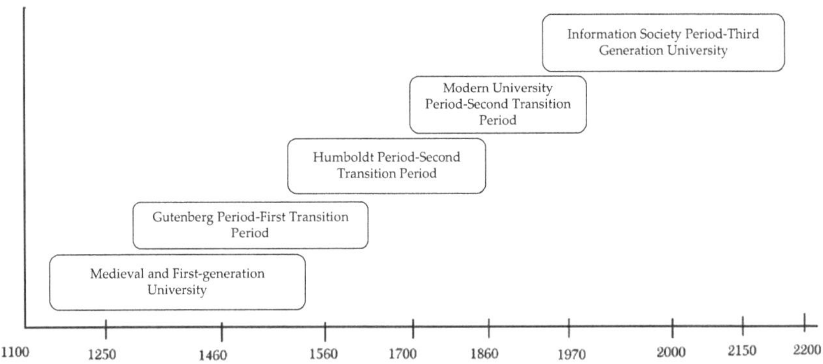

Figure 1.1 Universities' periods of historical change.

Source: Välimaa (2014) and Wissema (2009).

Characteristics of universities	First-generation	Second-generation	Third-generation
Objective	Education	Education plus research	Education and research plus know-how exploitation
Role	Defending the truth	Discoverng nature	Creating value
Method	Stochastic	Modern science, mono disciplinary	Modern science, interdisciplinary
Creating	Professionals	Professionals plus scientists	Professionals and scientists plus entrepreneurs
Orientation Language Organization	Universal Latin Nations, Faculties, Collegas	National National Languages Faculties	Global English University institutes
Management	Chancellor	Part-time academics	Professional management

Figure 1.2 Characteristics of universities over the three generations.

Source: Wissema (2009).

globalization (Vaira 2004). Therefore, in countries where neoliberal policies are dominant, the share of the state in HE is decreasing. In such states, companies have started to take the copyrights of education programs and services by collaborating with HE institutions to raise the flexible and well-educated employees and technology-acumen consumers they need. On the other hand, HE institutions have attempted to look for ways to obtain financing from alternative sources to compensate for the decrease in state support. This situation brought issues such as regulatory institutions, accountability, transparency, performance, and quality assurance in the field of HE institutions. Daniel (2015) summarizes all these developments in the field of HE under the four major headings as economic, technological, social, and educational, and Figure 1.3 indicates the global trends affecting HE institutions.

The major development trends summarized by Daniel (2015) as in Figure 1.3 have caused countries around the world to review and restructure their HE systems. George (2006) has revealed that there are two basic models, namely the neoliberal model, and the state-centered model for changing and developing HE systems. According to George (2006), the neoliberal model is based on the theory developed by World Bank theorists and practical examples of the USA, UK, Australia, and New Zealand, and focuses on creating a market where HE institutions can compete with each other by reducing the role of governments in HE (George 2006). The state-centered model expresses state investments and policies that will ensure that HE institutions are closely connected with developing industries, and is concerned with the characteristics of the countries that adopt the Asian

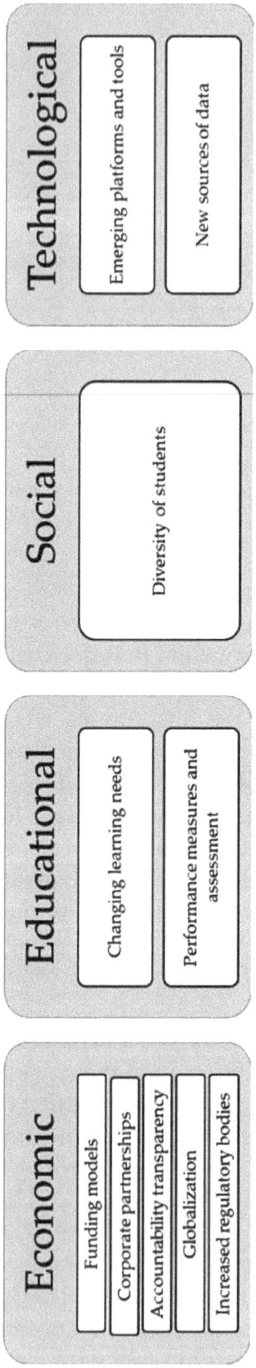

Figure 1.3 Major developments in higher education.

Source: Daniel (2015).

economic development model and the comprehensive analysis of European higher education models and many countries of the former Soviet Union. George (2006) defines higher education systems under two headings, namely, the neoliberal model and the state-centered model. The author also points out that these two models can coexist in a country. According to this approach, on the one hand, a higher education system may be closer to a specific model in general, on the other hand, as some of the issues such as financing, management, and curriculum, etc., can be based on the other model. For example, in a public university, administrative issues such as the appointment of faculty members are handled by the state (state-centered model), while the same public university may be under the sponsorship (neoliberal model) of an independent accreditation committee (George 2006).

Whether it is closer to the state-centered model or proceeding with a completely neoliberal model, the shape and content of academic activities have started to change. Changing academic activities also differentiated the management approach of higher education institutions. Changing and differentiating academic operations and management style has transformed HE institutions into business-like organizations, enabling business-objects tools and practices such as performance management and business analytics to be used in HE. These processes of change and transformation in the field of HE is explained in the literature with different theories, approaches, and concepts such as academic entrepreneurship, academic marketization, and new managerialism (Slaughter and Leslie 1997; Deem 2001; Pomeroy 2014).

1.1.1 Knowledge-based economy, academic entrepreneurship, and new managerialism

The primary needs and expectations from the organization and governance of universities have changed over the last few decades. There are two broad sets of ideas about university governance how organizational and decision-making structures within universities are established. According to the first proposition, the university is a "republic of scholars" whereas, in the second one, the university is treated as a "stakeholder organization" (Bleiklie and Kogan 2006). In this respect, these authors argue that emerging knowledge regimes can be divided into two major groups, namely academic capitalist regimes versus public managerialism regimes. Institutional autonomy and academic freedom are taken into consideration in the first idea meant that the leadership and the decision-making processes in the universities are based on collegial decisions made by independent scholars. In the second idea, institutional autonomy is considered a basis for strategic decision-making processes. In this case, the leaders are assumed to see it as their ultimate role to assure the interests of major stakeholders and where the voice of academics within the HE institutions is but one among several stakeholders.

In that sense, new managerialism is introduced by the private sector "solution" to the public sector "problem" in the HE industry (Milliken and Colohan 2004).

As a result of the KBE, universities' entrepreneurial activities and studies for the commercialization of research activities caused HE institutions to display market-oriented and/or profit-oriented behaviors like businesses. Market-like behaviors that are beginning to appear in universities and faculties are defined in the literature with the concept of "academic marketization" (Slaughter and Leslie 1997). The academic marketization brings the vision of the KBE into the HE industry and this vision leads to the transformative influence in higher education and research (HER) in the long term. One consequence of these changes is increasing criticism of education for lacking the human capital needs of a supposedly ever more competitive, extensive KBE (Jessop 2017). Horta (2009) argues that one of the reasons for this rapid change is related to the changing strategy in funding for research by enabling universities to retain and authorize the intellectual property rights considering their discoveries and inventions.

It is a fact that similar progress has occurred in various countries. Some of them are to some extent, coordinated through international agencies such as the OECD, or to a limited extent, emerging independently in response to the same competitive enforcements and the relative power of the KBE as a response to disruptive technologies and economic crises. There is a related trend, not only in the HE industry which is intensified global competition for talent regarding students at undergraduate and master level, and doctoral and post-doctoral researchers, but also, the highly skilled knowledge workers, and effective entrepreneurs (Salem 2014). This trend is the most obvious in science, technology, engineering, and mathematics (STEM) subjects. However, it is penetrated in the social sciences and even the arts and humanities when considered relevant to competitiveness in the innovative, cultural, and royalty approaches. Hence, instead of STEM, OECD (2013) recommends STEAM framework for satisfying future innovation and digitalization trends by including "A-Arts" to the list. Particularly, it is the top priority of government agendas to achieve social control. It is still unevenly built up, and while many HE institutions are called to join in, fewer can respond to this approach effectively. On the other hand, Marginson (2013) and Hemsley-Brown (2011) state that the capitalist visions cannot be entirely accepted in HE. There are some recent OECD surveys related to academic marketization and these surveys on universities clearly express these stresses indicating the difficulty in switching uneasily between their role in providing public goods or private benefit for students and other stakeholders in the HE sectors.

A major characteristic of this century is the "knowledge economy" (World Bank 2012; Thomas and Chopra 2020). Thus, there is a significant need for knowledge creation and dependence on knowledge, and intellectual capital (IC). The "knowledge pyramid" which is used for explaining the conversion process of data–information-knowledge–wisdom, contains a great opportunity for HE sectors in the globe (Pauleen and Wang 2017). Especially, the emergence of advanced technologies, intensive usage, and reliance on the internet and smartphones, and the arrival of tech giants like Apple, Amazon, Facebook, etc., has multiplied massive amounts of data. While classical information systems were collecting data and processing it to produce information, technology trends made it possible to develop new ways and means of analyzing and utilizing this exponentially increasing amount of data. That creates the knowledge to satisfy the needs of individuals, firms, and all types of decision-makers in an increasingly competitive world by better and faster decision-making. The HE sector is not an exception. It also has a second mission to support producing knowledge and sustainable digital transformation utilizing developing new technologies, methods through research and teaching. In this context, it should be noted that this digital transformation need is restricted to neither specific departments nor specific sectors, i.e., it is a global issue (Sumbal et al. 2017). It has a cross-functional nature and is related to various aspects of different sectors (Andreeva and Kianto 2012).

Digitalization has made real-world analysis fast, easy, and less costly for the business world. Likewise, such a robust KM approach with the emerging big data, technology shapes the framework and business model of HE sectors. In this respect, HE institutions are required to identify the relationship[3] between these big data and KM. There is a strong connection between them, i.e., value-added decision-making is based on the processing of big data through a knowledge-based intelligent system and the usage of KM is diverse, making it more sophisticated in practice (Pee and Kankanhalli 2009; Crane and Self 2014; Sumbal et al. 2017). Therefore, HE sectors have the key role to determine the results of KM in various aspects.[4]

Big data has been treated as a starting point for the knowledge generation process. In this sense, it becomes challenging for organizations to collect relevant big data, digitize and digitalize it, and efficiently utilize it for the digital transformation process. To overcome this challenge, it is critical executing an integrated strategy. In this way, it will be easy to set up an effective communication system and establish digital processes to support the exchange of information and data analytics. Hence, this will create a holistic approach toward KM in the digitalization and digital transformation context and feed business culture regarding the knowledge-based decision-making within an organization (Figure 1.4).

An overview of interaction levels of big data structures with KM structures is given in Figure 1.4. The most crucial point here is the true understanding of the relationship between the two to achieve the targeted outcomes within the decision-making processes in the institutional structure and the implementation of them in practice shown in Figure 1.4. In this context, higher education institutions have vital duties. Especially, there is a need for competent and qualified human capital in the digital business

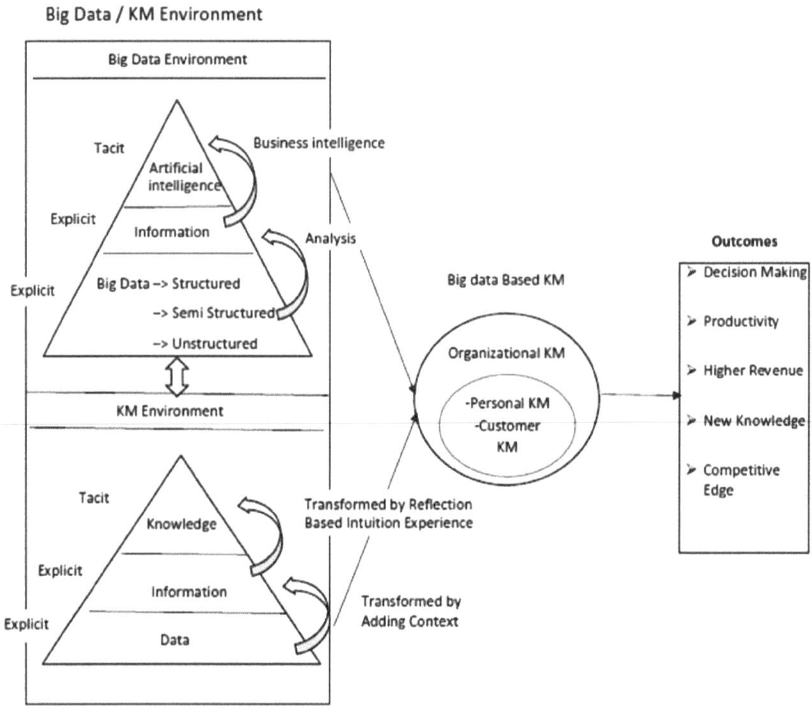

Figure 1.4 Relationship between big data environment and KM environment.

Source: Thomas and Chopra (2020).

world, and this issue emerges as a challenge. It is necessary for the educational content offered by universities to focus on raising professionals capable of transforming the innovations and technological developments at a more advanced level to implement their basic strategies in this direction.

1.1.2 Beyond R&D in the universities: Knowledge-based capital for global value chain

A mechanism that produces its means of production can be expressed as "knowledge-based capital (KBC)". This mechanism is also critical to producing the knowledge required in the future. This knowledge will become a national and global "public good" in the future. The most important example of this can be expressed as the approaches that the developments in the field of sustainability have emerged in national and global markets. The most important source of this new knowledge generation approach, which will come together by many elements, will be the technology transfer offices, institutes, and research centers within the universities. In this process, digitalization will enable many activities that take place within the body of universities and to become independent from the physical place. In this respect, the digital business processes of those involved in KBC generally consist of a "start-maintain-finish" cycle and will be fast. In this framework, the new management approach will be based on the joint work of individuals, regardless of their locations (Abel and Deitz 2010). Considering today's e-business environment, the headquarters of especially sizeable companies are in one country, while their employees are located in different countries and work continuously at different time intervals. This means that, together with digitalization, e-business processes that are served 24 hours a day arise at a high speed.

With digitalization, new business areas and new competencies are needed. In this context, it is known that many companies have difficulties in finding employees who are competent in the labor market and will adapt to the changes in the economy and the digitalized business world. A study by OECD (2012, 2013) provides recommendations for developing and managing a competence strategy to balance the supply and demand of talent and be more flexible. OECD policy recommendations about KBC to achieve "Global Value Chain *(GVC)*" are summarized as follows (OECD 2013: 64–65):

1. Incentive to learn throughout life
2. Incentive to fill the gaps by international mobility of skilled people
3. Incentive to invest in cross-border higher education
4. Incentive to activate workers by financial support
5. Incentive to prevent early retirements
6. Incentive for entrepreneurship

With Education 4.0, universities will turn into institutions that do not possess a spatial dimension for KBC production. There will be a need for visionary universities that produce and develop Industry 4.0 infrastructure in the said process (Malseva 2018). However, problems may arise in the control and risk management processes of such structures that differentiate over time and grow in scale. As a solution to these problems, internal audit and internal control have become fundamental management techniques.

Just as digitalization requires the continuous improvement of program algorithms and software programs, a structure that requires ongoing monitoring of digitalized business processes emerges. That is called "continuous auditing and continuous monitoring (CA-CM)" (Vasarhelyi and Halper 1991; Vasarhelyi 2002; Bozkus Kahyaoglu et al. 2020). In this sense, replacing a loss in business processes increases costs. Especially when we consider the structure of universities, the management style in the HE sectors where different disciplines and branches come together should be based on algorithms. Therefore, a risk-based audit approach for proactive, early detection and prevention of risks is needed instead of reactive fraud and investigation activities (Vasarhelyi et al. 2004; Vasarhelyi and Kuenkaikaew 2010). This requires an approach to apply CA-CM techniques to minimize risks in the digital environment. Thus, internal audit becomes a mechanism for achieving strategic goals, considering the "purpose and means" balance in the institutional functioning. The critical point here is that universities have both economic and social goals in the education process (Trequattrini et al. 2015). Auditing is an important tool to determine whether it meets the needs in achieving these goals and to evaluate its effectiveness (Vasarhelyi et al. 2012; Vasarhelyi et al. 2015). For this reason, auditing in the universities, not only provides reasonable assurance in economic matters, but also added value in "social auditing[5]" (Pareek 2014).

1.1.3 Digital strategy and e-business model for Education 4.0

Digital strategy can be defined as "the integration of digital technologies into organizations' strategies and operations" in such a way that radically transforms the value chain (Sambamurthy and Zmud 2017). In this respect, the digital strategy concept has emerged as an outstanding source of "competitive advantage" by being a locomotive of revolutionary changes in all kinds of products and services HE institutions generate for the related market segments and hence, how they perform their businesses in the education sector. It is a fact that "knowledge" is at the heart of this digital transformation, i.e., tools to design it, use it, and interface it. Cevik Onar et al. (2018) state that there is a strong link between Industry 4.0 and Education 4.0 and advocate that education in universities should reach a level that can meet the needs and expectations of Industry 4.0.

University management, including internal auditors, need to assess the major business impacts of the digital transformation touches each case on the value chain from front-end learning and academic experiences to back-end operations, support systems, and supply chain management. It should be noted that in certain cases, they may affect the total business model transformation. In this context, these new tools for assessing the new approaches and learning from experience and knowledge are significantly transforming the way critical decision-making process is made across in a university. However, internal auditors should remind the university's top management that just because technologies are digital does not necessarily mean that they obtain any specific outcomes without applying a focused management style. In other words, HE institutions are recommended:

1. To use new organizational models to maximize tactical maneuverability
2. To establish a clear vision around digital strategy and transformation across the management team[6]
3. To optimize the digital governance models

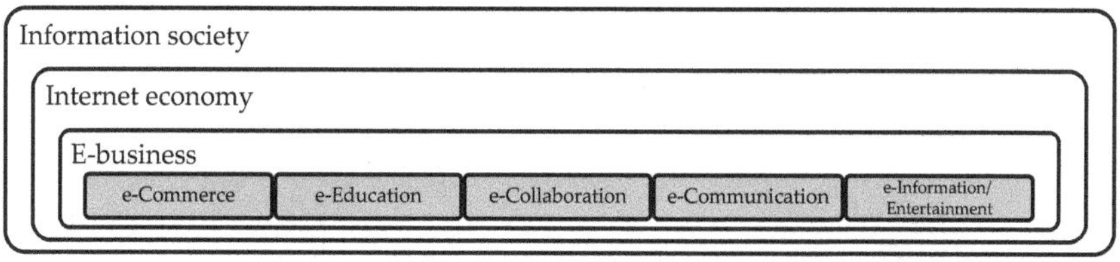

Figure 1.5 Multidimensional approach for Information society and Internet economy.

Source: Wirtz (2018).

The impact of digital transformation can be defined as a "digital revolution or Education 4.0[7]" and in this respect, the digital revolution in the education sector occurs in widespread, i.e., it is touching not only the core business but also leading to social and political changes all over the world (Denger and Wirtz 1995). According to Schumpeter (1939), innovation is determined by "societal development" and this approach is based on the *Kondratieff theory* that estimates the technological innovations through the sinusoidal innovation phases. In addition, Wirtz (2018) states that the dynamic digital development process is not a new concept, rather the development of information and communication applications goes back to a long history. Information society and the Internet economy are established by multidimensional elements of e-business[8] namely, "e-Commerce", "e-Education", "e-Collaboration", "e-Communication", "e-Information/ Entertainment", which is shown in Figure 1.5.

E-education definition is as follows: "the transfer of education and training services to third parties utilizing electronic networks". The major aim of e-education is to deliver these services in a resource-efficient manner by using location and time-independent applications of electronic networks (Turban et al. 2015). Although the digital transformation process is still in its early stages of development in the universities, two distinctive underlying schools of thought emerged on the best model for achieving this. The first one claims that the digital capabilities of HE institutions need to be integrated into the current business model. The other one asserts that the dedicated digital best practices need to be carved out with different enforcement, hence, different integration competencies (Wirtz and Daiser 2017). Figure 1.6 provides a summary of these critical issues. Some key questions are starting with "what?", and "how?" to decide on the best digital transformation strategy and Education 4.0.

It is a fact that the digital transformation in HE institutions is the outcome of the three major elements that have both external and internal roots (Wirtz and Daiser 2017; Sambamurthy and Zmud 2017):

1. New facilities and resources
2. New technologies and knowledge for transformation and allocation
3. New mass markets for academic teaching

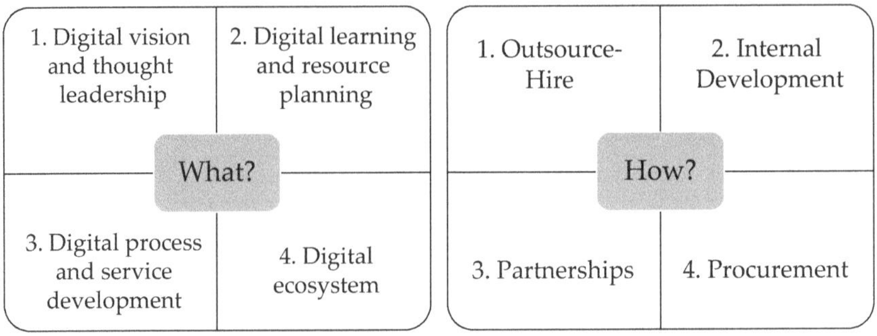

Figure 1.6 Key Issues for digital transformation strategy development and execution for Education 4.0.

Source: Wirtz and Daiser (2017).

In time, these inputs lead to the progress of new products and services that convert end-user behavior, modern infrastructure to fit in with those products and services, and advanced management systems and organizational models (Figure 1.7). HE institutions are exploring new approaches for arranging their operations to obtain the benefits of the new technologies without sacrificing internal control, cybersecurity, and risk management

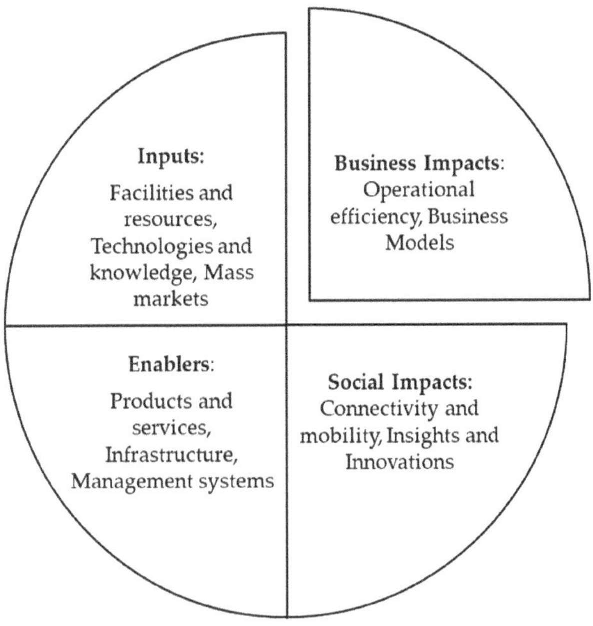

Figure 1.7 Key elements of digital transformation in universities.

Source: Sambamurthy and Zmud (2017).

activities. Thus, HE institutions are not only altering their roles, responsibilities, and decision-making processes but also generating a vision supported by strong leadership and governance systems designed to allow the execution of digital transformation strategies.

The digital transformation has increased the quantity and velocity of information through corporate communication channels. Hence, this leads to meld horizontal communication channels on top of the traditional vertical ones and transfers strategic decision-making power to the front lines. Some HE institutions are taking further steps in evolving their organizational structures. They are changing their organizational structures from the traditional unified designs ("U-Form") in favor of multidivisional arrangements ("Digital-Form") to gain more improved coordination and decision-making processes to manage more diversified and complex business models in the new education sector shaped by the digital landscape (Figure 1.8).

The business model framework was first used in the literature in the 1990s. The point to be noted here is that the concept of business model is defined mainly by associating it with "computer-assisted information processing", "computer system models", and, "computerized models" (Wirtz 2019). Hence, the conceptual framework of the business models is generated from the concept of information modeling based on the way for the production process of information. As a part of the evolutionary process of mankind, institutions are all designed and established by humans. On the other hand, cognitive processes like perception, recognition, decision, etc., are rapidly changing by using new and advanced technologies such as visual analytics, augmented reality, and decision support systems, etc. Advanced technologies and new tools such as big data, the Internet of Things (IoT), communication channels and networks, augmented reality, sensors, social media, etc., are embedded in this infrastructure. They are all becoming a required part of it

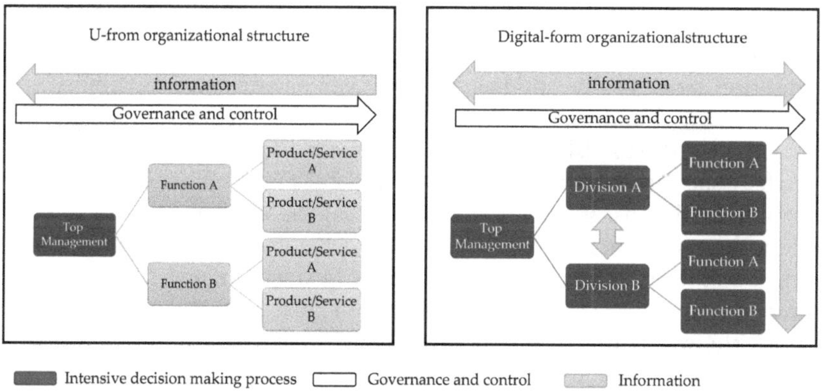

Figure 1.8 Digital DNA for organizational forms.

Source: Stone (2019).

and hence, they have the power to change the perception of the organization. Kahneman (2011), the Nobel Laureate in Economics, defines this issue as "the Subconscious Minds of Organizations" and the author introduces the subconscious and consciousness based on behavioral psychology to the decision-making processes in the economic sector. There is a huge expansion of institutions' "conscious and subconscious mind" by means of the increasing connectedness which is enabled by these sophisticated and highly automated tools. Artificial Intelligence (AI) implementations are embedded in a set of structural and technological systems on an institution thus, forming a "subconscious mind of organizations" which interacts with the conscious activities performed by the institution as a whole or business units (Kumaran et al. 2016). It is possible to design and influence the subconscious mind of the organizational units or organizations to establish a "digital business model or e-business model" by using new tools and advanced technologies. Therefore, it is critical to understand how to become and sustain innovative and thus successful with the e-business model.

Bradley et al. (2015) have an attractive perspective for the e-business model approach, and these authors define the value of Digital Business Models in three categories, namely, cost value, experience value, and platform value, for the organizations. When this approach is applied to the HE sector, the cost value will be related to transparency in tuition fees, consumption-based pricing, rebates, and rewards. Secondly, the experience value is related to students' choice, personalization, automation, lower latency, and any device at any time. And thirdly, the platform value is related to online education platforms, crowdsourcing, peer-to-peer, sharing knowledge and economy, and data monetization.

Wirtz et al. (2016) define the e-business model classification based on individual subcategories shown in Figure 1.9. These components in Figure 1.9 constitute a so-called part of the integrated business model approach. In other words, the combination of those core elements ensures together a functional and integrated business model. This e-business model classification with the conceptual framework is crucial to represent how an institution creates value and hence, how it can ensure its sustainability.

Both internal and external conditions of an institution should be taken into account to establish a relevant e-business model. In that sense, industry-specific issues are considered in environmental factors as a leading part of external conditions. Wirtz (2019) argues that the configuration of the partial models of an integrated e-business model is important to achieve high-performing institutions. Each integrated business model is formed by different partial models and this is shown in Figure 1.10. According to Wirtz (2019), there are three major components with nine partial models within a typical e-business model. These three major components are the strategic component, stakeholder component, and value-added component, respectively. Since this book specifically focuses on the HE industry, the e-business model components are explained in this context in the below sections.

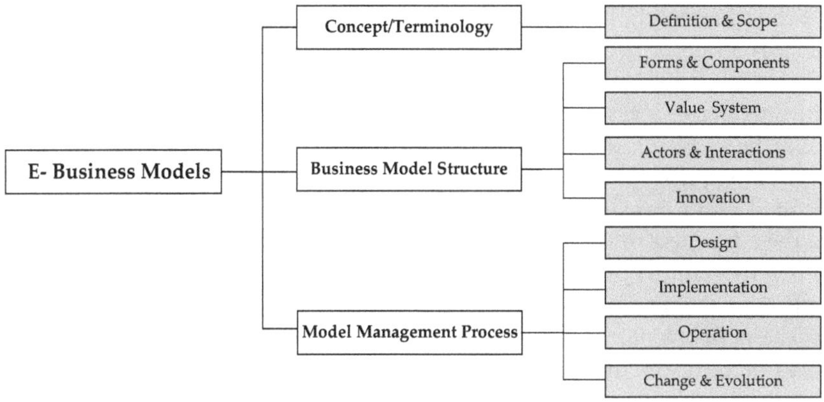

Figure 1.9 e-Business model classification.

Source: Wirtz (2019).

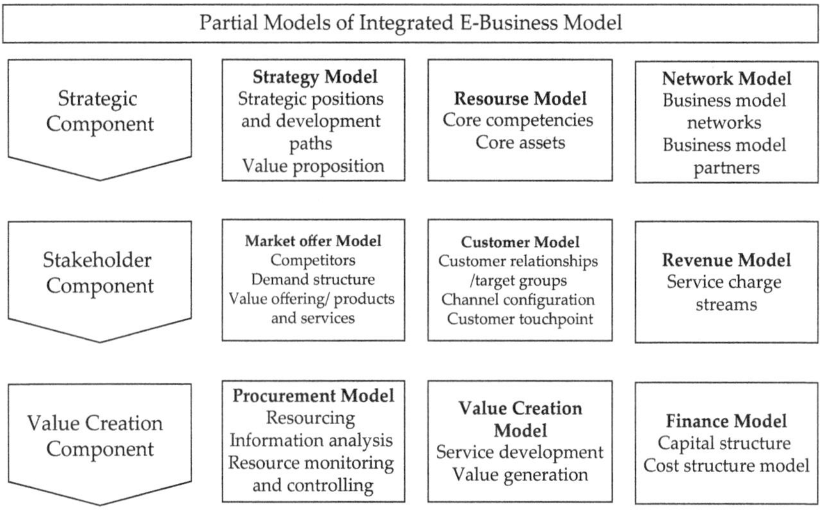

Figure 1.10 Partial models of the integrated business model.

Source: Wirtz (2019) and Yang et al. (2014).

The first one is the *strategic component* of an integrated e-business model and it consists of three partial models, namely, the strategy, resource, and network models, respectively. In the *strategy model*, the top management, i.e., rector and faculty deans state the goals and activities of the university in order to sustain in a highly competitive education sector. These strategies should unite the academic vision, mission, and goals. In this context, positioning the university and defining the strategic business areas such as teaching, research, and development, etc., is closely related to this component. A strategic situational assessment is necessary to shape the changes in framework terms; the scope of measures and strengths and weaknesses of the university serves as a basis.

In the *resources model*, the core assets and core competencies are indicated along with their related factors relevant to value creation. They can be defined as a summary of all suitable, tangible, and intangible input factors of the e-business model applicable for the HE institutes. During this process, it is important to present both internal and external resources and competencies of the university staff in detail.

The *network model* provides an outline of the value-generating partners in value creation and the relationship between diverse business models. In this respect, the network model becomes a tool of the university's top management to monitor and guide the value allocation within a collaborative value creation process. In this regard, it is recommended to analyze both tangible and intangible streams of information and digital services. In this way, stakeholders of the university in value creation may be identified and ranked based on a network of connections and relations.

The second one is the *stakeholder component* of an integrated e-business model and it consists of three partial models, namely, the market offer, customer, and revenue models, respectively. The stakeholder component and its partial models indicate the primary impact items for producing the style and conducting the operation of a business model. These partial models provide valuable information about the corporate culture and business environment of the education sector-specific for HE institutions. In addition, they are used for connecting the corporate culture and revenue generation process of universities via internal value creation mechanisms in the education sector. In other words, they can be treated as the indicator of a link between the corporate strategy and its value-generation process. It is a fact that the internal value-generating process of a university should be determined by understanding the market conditions such as demand structure, and competitors to make the best offer to the target customer segments. In this way, the university's "value proposition" is decided and shaped to gain a competitive advantage against the background of an analysis of competing business models in the education sector (DeWit and Meyer 2010; Coaldrake and Stedman 2016). Hence, this approach is defined as a "market offer model" and aims to exploit the available market opportunities. The customer model involves all kinds of activities in the HE institutions aiming to satisfy the needs and expectations of the students. Therefore, similar to the other sectors, the major focus in the education sector is to establish an effective customer relationship management (CRM) and to maintain a loyal target group orientation. This is also the major target regarding the generation and design of marketing and promotion channels. In this respect, HE institutions are investing more and more in e-business platforms to capture young generations as their key customers. They understand the importance of "student orientation" which is crucial regarding the setup of customer interfaces – in other words, "customer touchpoints" – which are the sole points of interaction between the university and the students and researchers (Wirtz and Daiser 2017). In the *revenue model*, the HE institutions determine the relevance and actual revenue streams of their revenues based on their corporate strategies and business models, respectively. In this context, universities measure their success levels regarding the value capture of internal value creation. In other words, the revenue model is used as a tool to monitor "how and to what extent the value gained" can be monetized. The revenue model is thus accountable for the effectiveness and efficiency of the value-generation process of universities.

The third one is the *value creation component* of an integrated e-business model and it consists of three partial models, namely, the procurement, value creation, and finance models, respectively. The value creation component contains the internal value-generation process for the education sector. The value creation process generally requires the functionality of the corporate structure (Baden-Fuller and Morgan 2010). In this context, the connection and interaction of this partial model with other parts should also be considered.

The partial model of *procurement* is essential for the education sector regarding the "right" resourcing, resource monitoring, and controlling process. In this model, information analysis is the key to success, and digitalization is required to reach a strong corporate culture. The procurement model defines the administration and origin of the materials, which are essential for the generation of value-added goods and services for the universities. In addition, *the value creation model* and *finance model* are closely related to the procurement model. In this case, the HE institutions must establish a relevant e-business model to cover the cost structure and capital structure for a sustainable growth strategy (Amit and Zott 2012). From this point of view, universities' success in one or more of the above-mentioned partial models does not bring the expected outcome (Beheshti and Salehi-Sangari 2007). Therefore, it is foremost to consider the e-business model altogether and make decisions accordingly while determining their corporate digitalization strategies to achieve Education 4.0.

1.2 RETHINKING THE SOCIAL CONTEXT FOR DIGITAL TRANSFORMATION IN THE HIGHER EDUCATION SECTOR

According to the report of the World Bank Independent Evaluation Group (IEG) (2017), the Higher Education sector has multiple roles which go beyond just educating learners. These multiple roles are defined with interrelated missions (Salmi 2009; Sánchez-Barrioluengo 2013; Cleary and Van Noy 2014):

(i) "Teaching and learning" to better prepare learners for the future
(ii) "Research" to better spillover innovation and knowledge
(iii) "Community engagement" to better collaborate with university and public–private partnerships

It is a fact that HE sectors around the world face a rapidly increasing demand for enrollment that is characterized by non-traditional learners. In this respect, limited capacity to respond to such demand among existing state-owned universities, in turn, has led to considerable growth in private arrangements in the HE sectors. In this respect, limited capacity to respond to such demand among existing state-owned universities, in turn, has led to considerable growth in private arrangements in the HE sectors. In addition, there is an increasing demand for skilled professionals and researchers to satisfy learners' needs and expectations. In this context, governments are required to develop policy instruments to create social benefits in the HE sectors. These requirements are summarized as (i) the increasing demand for the HE sector; (ii) the access to the HE sector; (iii) the quality standards of the HE sector; (iv) the research and development activities in the HE sector; and (v) the reform needs of the HE sector in a digital era (Martin and Parikh 2017).

It is accepted that investing in HE sectors improves the capacity of a countries' competitiveness (World Bank 2011, 2016; Sala-i-Martín et al. 2015). That may bring some distribution consequences. Especially when the HE is executed traditionally, higher education tends to benefit disproportionately the advantaged groups. On the one hand, there is an increasing demand for a more educated labor force; on the other hand, there are some concerns about the quality and compatibility of HE, and about value for money and resource allocation in the HE sector.

Digital transformation in the HE sector is mostly about rethinking the organizational culture, the code of conducting teaching and research, providing services in a high-tech, and being sustainable in a digital ecosystem. Faculties, research centers, and related administrative departments must support the adaptation of the digital-first response and culture, while IT departments must be more alert to be skilled to align with the new technology speed and educational policy outcomes. In this respect, there is a need for a collective and shared "digital delivery model" created by the joined working ethos, where faculties and IT departments work together (Nissen 2018). This approach may bring an opportunity to reinforce collaboration and synergies with other HE institutions. Indeed, the digital transformation of the HE sector is not just an ordinary technical recurrence. It transforms how the HE institutions are organized, how critical decisions are made, and how educational policies are determined and enforced. Digital transformation will be most likely to bring a disruptive change that requires strong leadership to manage. It will also need strong communication through digital means, both internally and externally. It will also trigger multiple digital risks (Buckingham Shum and Ferguson 2012; Picciano 2014). Since digital transformation is a fundamental requirement, it must be applied irreversibly without leaving it to people's choices.

According to Jensen (2019), a Manager for Technology in Higher Education at the International Association of Universities[9] (IAU), digital transformation is seen as a shared priority by 68%, and leadership support by 72% of HE participants on a recent global survey conducted by IAU. Although there is a high level of awareness and leadership support for the digital transformation, there are still challenges and common issues to be addressed. This situation is directly related to the social context of digital transformation. Hence, there is a need for rethinking the social context of digital transformation in the HE sectors.

It is a fact that learning is a social activity. In this respect, the individual learner's willingness and ability to sense each other's viewpoints is assumed to be depending on the ability to connect "socio-emotionally with collaborative partners" (Slavin 1990; Kulkarni et al. 2015). This situation gains more importance with the effect of technological innovations (Bizri et al. 2019). Considering the digital transformation in a social context, technological innovations have two different impacts which are divided into two broad categories namely, evolutionary[10] innovations and revolutionary[11] innovations, respectively. Kiss and Vass (2019) have developed an approach on how these two definitions of innovations should be interpreted for the HE sectors. Firstly, they claim that "implementation of knowledge, vision, and creativity" are central factors of the digital transformation in the HE sectors. They propose new concepts and novel ideas as the starting site of innovation in the HE sectors. Secondly, they argue that students' and lecturers' needs and expectations should be satisfied. In this respect, sharing the needs and expectations before starting curriculum mapping is a strategic goal for HE institutions. That is the substantial case for diagnostic assessment during the digital transformation process based on educational data analytics. Thirdly, the authors state that the two major categories of innovations are strongly determining the shape of the higher education sector. Particularly, revolutionary innovations are seen in the processes of HE institutions. The authors argue that when there is a "conscious coherency" among the key concepts, namely innovation, change, culture, curriculum planning, teaching methodology, and assessment, then this can produce continuous or dynamic evolutionary innovation in HE institutions (Kiss and Vass 2019: 325–326).

There is a need for mutual awareness and understanding at both local and global levels to solve issues. The sustainable digital transformation of the HE sector is the key to a value-added approach for shaping a relevant, human-centered digital future for the collective global goodness, regardless of wherever we live in the world.

1.3 TRANSFORMATION OF A UNIVERSITY: THE CASE OF MEGA-, MACRO-, MESO-, AND MICRO-LEVEL ANALYTICS

Higher education institutions are under pressure to satisfy the growing social demands based on global digital transformation (Daniel 2015; Nguyen et al. 2017; Kiss and Vass 2019). In this respect, it is challenging for academic staff, especially for the researchers to react effectively and in time to the social demands and global changes occurring in the HE sectors. Considering the advances in the technology and big data generation process in the last decade, it is not surprising to witness rapid development in the field of educational data analytics as well. For this reason, educational data analytics is highly recommended by various authors for coping with these challenges and pressures associated with determining the proper information at the right time to assist the decision-making process at the institutional level (Baker 2010; Siemens and Long 2011; Baker and Inventado 2014; Daniel 2015; Sin and Muthu 2015; Nistor and Hernández-Garcíac 2018; Nguyen et al. 2020). It is a fact that the major HE institutions have invested significantly in recent years to either acquire or develop educational data analytics tools. In addition to the relevance of educational data analytics tools and technologies to meeting regulatory concerns of HE institutions, preceding works have also demonstrated that they are used to achieve the commercial goals of universities themselves.

Daniel (2015) claims that educational data analytics can be a functional tool for monitoring learning outcomes and increased quality standards of three pivotal end-users in the HE sectors; namely students, administrators, and lecturers. Figure 1.11 indicates the key end-users, and the scope of educational data analytics, respectively. Concerning these end-users and the extent of analytics, the fundamental educational data types. Curriculum data, Administrative data, Department data, Teaching data & Learning data, Research data, and Student data are given in Figure 1.12.

In Daniel's (2015) work, it is not only the benefits of educational data analytics in discussion but also the key risks are described regarding the ethical issues, the quality of data, cybersecurity, ownership, and privacy issues. Similarly, Slade and Prinsloo (2013) and Jones (2012) state the challenges and ethical dilemmas about institutional responsibilities related to data access relevant for the HE sector based on the outcome of these analytics. As a possible solution to these issues, Dringus (2012) and Dyckhoff et al. (2012) propose to use the transparency and flexibility of learning analytics to deter any potentially improper access of data.

It is a fact that in the current educational systems, including "Learning Management Systems (LMS)" and "Course Authoring Systems (CAS)". There is plenty of information gathered as big datasets originating from routine daily operations in the HE sectors (Figure 1.11). Data analytics in HE is essential for controlling, managing, and modeling educational

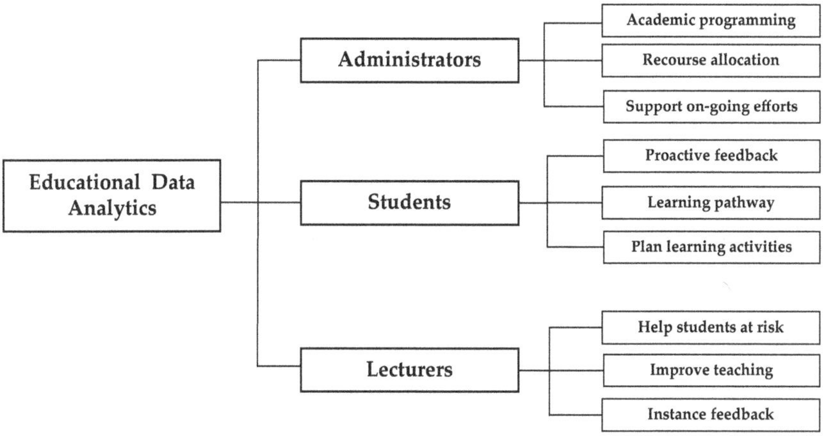

Figure 1.11 Key end-users of educational data analytics.

Source: Daniel (2015).

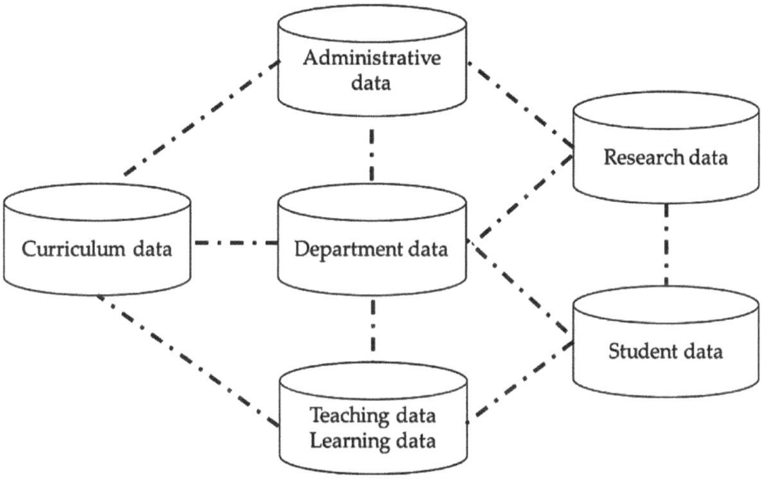

Figure 1.12 Basic types of educational data.

Source: Daniel and Butson (2013).

processes (Romero and Ventura 2010; Greller and Drachsler 2012; Dahlstrom et al. 2014; Chaurasia et al. 2018). In this respect, three major methodologies and processes are promoted for data analytics in higher education, namely "Academic Analytics[12] (AA)", "Educational Data Mining[13] (EDM)", and "Learning Analytics[14] (LA)". Although these data analytics methodologies are relatively new fields in the literature, they are closely linked to each other. In other words, they provide valuable information where the result of one methodology can be used as the input of another both in teaching and learning environments (Siemens 2013; Zouaq et al. 2013; Dahlstrom et al. 2014; Peña-Ayala 2014; Avella et al. 2016; Nguyen et al. 2020) (Figure 1.11).

On the other hand, van Barneveld et al. (2012) argue that there is room for improvement to achieve the consensus of the terms in these fields. In this respect, the work of Nguyen et al. (2020) is an attempt to fill this gap by providing an integrated approach for educational data analytics. In addition, Ifenthaler (2015) proposes a novel perspective, for learning analytics considering the needs and expectations of diverse stakeholder levels in HE sectors. This approach shows the educational data flow between stakeholders in HE sectors (Figure 1.13). In this way, it is possible to monitor the condition of LA at each level regarding the learning context.

According to Ifenthaler's (2015) approach, the first level where the learning activities occur is defined as the "micro-level", where the learners are close to the learning environment. The micro-level analyses include one-on-one interactions between relevant parties. Analyzing at the micro-level implicitly encapsulates the behavior of individuals: (a) Their participation in and dependency on HE sectors, (b) Their access to HE sectors, and (c) Their power of investing in their education. Briefly, the micro-level analysis focuses mainly on individual engagements and dependency in HE sectors. The inquiry studying the learner groups and the interaction within these groups is described as the "meso-level". The meso-level analysis of HE sectors focuses mainly on understanding how HE institutions function in terms of structural characteristics, interactions between learner groups, and the implications for supply, demand, and tuition fees. HE sectors functioning

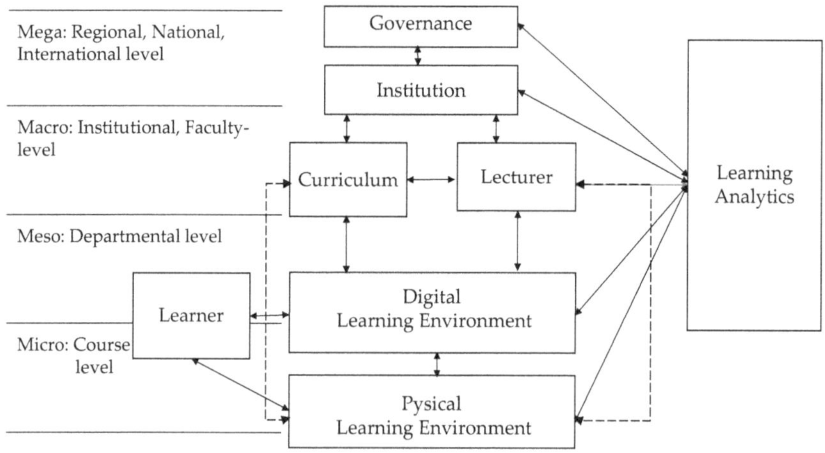

Figure 1.13 Educational data analytics with stakeholder levels.

Source: Ifenthaler (2015).

represents the best practice of the organizational structure, conduct, and performance of faculties and research institutions.

The third level defined as the "macro-level", contains effective information for system designers, researchers, and academic staff with the curriculum and learning design elements essential for educational data analytics. In this way, the design decision-making process provides visionary information at the macro-level facilitating educational data analytics to measure the effectiveness and efficiency of operational processes, and resource allocation at the entity level.

The fourth level defined as "mega-level" indicates the supreme level of the LA framework designed for incorporating data from all lower levels. In this way, it is possible to make benchmarking in the HE sectors. Thus, mega-level analytics is a tool to compare governance models with practical understanding through the recognition and documentation of instances within and across the HE institutions. Additionally, mega-level analytics contains predictive analytics and simulations that support university management for defining the appropriate educational policy-making (Manyika et al. 2011; Daniel 2015; Ifenthaler 2015; Nguyen et al. 2020).

On the one hand, HE institutions are the slowest-changing institutions in the digital transformation process; on the other hand, HE institutions are advancing as fast as they change. The main reason for this is that university staff can adapt slowly to this development. In this respect, a digitally transforming university approach has emerged to improve their staff through continuous education. Depending on the necessity of expertise in digitally transforming universities, the relationship and knowledge exchange between disciplines have increased notably. The most crucial issue is that the digital transformation occurring in every field reveals data that needs to be processed. Statistics and mathematics are the basic tools for analysis, and the necessity of processing the information with computers has made digital transformation the most decisive method of the knowledge-production process.

MacNeill and Johnston (2012) propose the "Conceptual Matrix for the Digital University" that is shown in Table 1.1. In this respect, Table 1.1 portrays the demand for a comprehensive, consistent outlook on "the digital", which is reaching beyond the typical model of technological infrastructure and implementations in teaching (Smyth et al. 2015: 13–16).

New techniques are put into effect with latest software in different disciplines. An evolution in physics can lead to a crucial development in the field of social sciences. This development has influenced universities through digital transformation in the knowledge-production process. The knowledge-production process turned to networks from the spatial dimension through digitalization and led to a transformation from the governance structure of the university to internet networks. If a university is defined as a place where knowledge is produced, centers like European Council for Nuclear Research[15] (CERN) (CERN 2017: 8–9) can be considered as universities too.

Table 1.1 Conceptual matrix for the digital university

Digital participation	**Information literacy**
• Globalization • Widening access • Civic role and responsibilities • Community engagement • Networks (human and digital) Technological affordances • Constructive alignment	• High-level concepts and perceptions influencing practice • Staff & student engagement and development • Effective development and use of infrastructure
Curriculum and course design	**Learning environment**
• Curriculum representations, course management, pedagogical innovation • Recruitment and marketing • Reporting, data, analytics	• Physical and digital • Pedagogical and social • Research and enquiry • Staff and resources

Source: MacNeill and Johnston (2012).

Based on the conceptual framework, there is a need for establishing a curriculum relevant to a digitally transforming university. Thus, digitally transforming universities will have a curriculum, which is different. The curriculum of digitally transforming universities is illustrated in Figure 1.14. Smyth et al. (2015) recommend a "curriculum which is located and co-located digital classrooms", including several linkages between students concerning the academic work of stakeholders and the extent to which this can be distributed beyond the university.

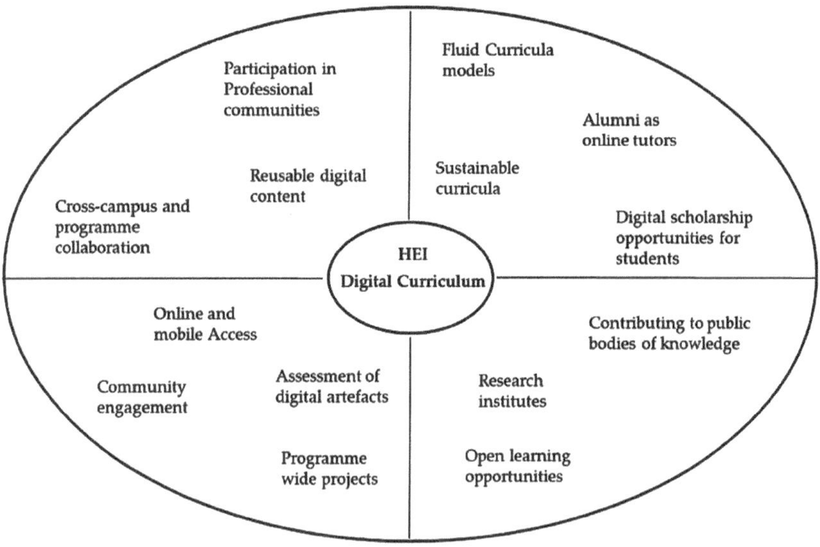

Figure 1.14 Curriculum of the digitally transforming university.

Source: DFWG (2014).

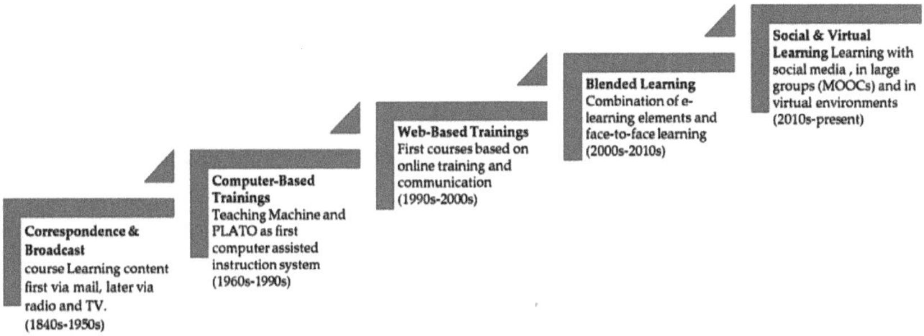

Figure 1.15 History of e-learning.

Source: Jeschke and Heinze (2014).

There is a change in the teaching and learning methodology all over the world. And after the pandemic, this change is taking place more rapidly, and even like a boom... (Jeschke and Heinze 2014; Arthur D. Little 2016). In this respect, it is possible to see how e-learning and distance learning gain importance in time based on the speed of the technology advancements by looking at the history of e-learning and Massive Open Online Courses[16] (MOOCs) as shown in Figure 1.15. Siemens (2012) defines learning in a creative way that is suitable for the realities of the virtual world such that

> Learning is "(...) focused on connecting specialized information sets (...) the connections that enable us to learn are more and more important than our current state of knowing. (...) Nurturing and maintaining connections is needed to facilitate continual learning. Ability to see connections between fields, ideas, and concepts is a core skill".

Needless to state, it is not enough to examine the future of universities and learning methods, but also the changing needs and expectations of society should be analyzed carefully since the way of living, working, and entertaining in society is also changing. Hence, this is called Society 4.0. Although emerging Society 4.0 manners involve some risks and uncertainties, Jeschke and Heinze (2014) argue that there may be opportunities for the HE sectors to establish a link between Education 4.0, Industry 4.0, and Society 4.0. With these opportunities, radical change and visionary perspective and stable leadership structure are a prerequisite. The below issues should be taken seriously by the university management and all stakeholders of HE sectors as well, to positively exploit these potential opportunities to achieve future university and learning context.

1. *Change in Institutional Structure*: There is a need for new business models of the universities to adopt the digitalization process. In this respect, faculty, department, and research center structures should be

renewed to be more cooperative and productive layouts via enhancing interdisciplinarity (Wissema 2009; Hulla et al. 2019).

2. *Change in Accreditation Structure*: It should be noted that examination styles in the universities will be changing due to radical changes in the teaching and learning environments. In this respect, examination offices in the universities will have a different role since fixed degree programs are disappearing soon. MOOC credits will be accepted as equivalent to their courses by various education providers, and a similar trend for the recognition process will occur in the analogy with EU Bologna Program (Arthur D. Little 2016; Cantú-Ortiz 2018).
3. *Change in Teaching Methodology*: There is a new teaching concept supported by new teaching infrastructures in the HE sectors (Wissema 2009; Faviero 2012). HE institutions will be taking it more seriously to recruit professors with better teaching skills and digital pedagogical experiences to be more competitive in the HE sectors.
4. *Change in Learning Structure:* There is no need for a classroom to learn new things in the future. And there is a "Personalized learning-do it yourself" concept available with new learning infrastructures e.g. increased computing capacities will drop down the costs of enrollment, virtual laboratories in humanities and social sciences, etc. (Rouhiainen 2019). In addition, there is a shift from presence learning to MOOCs all over the world (Arthur D. Little 2016). Instead of providing learning platforms, there will be distributed learning content available throughout the web. This approach, combined with credentials for distributed learning content, will provide an opportunity to gain faster accreditation cycles for new graduate programs in the HE sectors.

1.4 IMPORTANCE OF DIGITAL PARTICIPATION AND INFORMATION LITERACY FOR INTERNAL AUDITORS

The most important feature of HE institutions is that they have the function to classify and transfer information in a usable form and form a basis (Campbell and Carayannis 2013). This feature is the process of transfer of existing and past information stock to the business world. The pool of information here is "scientific knowledge". Those who know questions accelerate the application of scientific knowledge with the opportunities provided by technology in the HE sectors. This process also enables new analysis tools to be acquired. The function of HE institutions here is to provide the transfer of scientific knowledge in different fields interacting between distinct disciplines Cevik Onar et al. (2018). Specifically, the transition of development in the field of physics to social sciences with a conceptualization process and the transfer of a field such as psychology or neuroscience in medicine to computer sciences can be through universities. However, it is worth mentioning here that while technological advances accelerate the processing and application of existing scientific knowledge, they do not lead to the same development of scientific knowledge. The accelerating technology can be attractive in terms of the liquidity opportunities it has created within the body of large organizations. However, it is up to HE institutions to produce the basic information needed and make it available (Bernhard 2012).

It is necessary to know that the information called big data, brought out by the developments in communication technology, is not scientific knowledge. In this sense, universities gain importance as places where scientific knowledge comprises big data. Since the development process of technology creates a fast market, everyone can find a place in this realm. This phenomenon occurs predominantly via "digital participation" (Cevik Onar et al. 2018). In this respect, there is a mutual dependency. The only areas of cooperation for this interdependence are universities. The transformation of the information revealed by people from different fields and for separate purposes into a collective analysis tool will be a result of the new university understanding.

The critical point here is the necessity of teamwork and teams. Their facilitation can be done via digital participation. The need for deep specialization rising from scientific developments shows that the research and analysis should be done with extended teams. In this respect, the necessity of restructuring with an interdisciplinary understanding in all branches, especially in social sciences, stands out. Considering the risk of loss arising from projects involving people from different disciplines, the necessity of a new management approach comes to the fore. The most fundamental area of this management approach in risk management, internal control, and audit. In this sense, the new audit approach might be based on ethics (Jones 2012; Slade and Prinsloo 2013).

There is a distinct difference between big data produced in companies and big data produced in HE institutions. Although the knowledge generation process of universities is slow, their added value is higher. This added value is also because it is a place in the preparation process of new human capital. In this respect, universities are places where implicit knowledge is gained, especially for people and a resource that will increase output (Tapper and Palfreyman 2004). At this point, if the main purpose is to increase production, production factors should be defined in terms of content, although not conceptually. In this context, it would be appropriate to see HE institutions as a component of capital.

When all company structures, from space research to all other technological fields, since the 1990s, are analyzed, it is seen that there are people who invest in technology, and today there are people who are scientists in large universities. However, there is a transition from existing large companies to HE sectors. In this respect, it comes to the fore that all the information produced in new business processes is in a defined perspective under academic standards. It becomes a necessity for the transmission of knowledge. That necessitates the new university concept and trend. The greatest advantage that universities can provide for today will be reorganizing and bringing together those who use knowledge and information. This organization will be in a global area to the extent provided by digitalization regardless of location (Westerheijden and Kohoutek 2014).

1.5 CONTEXTUAL INFLUENCE AND RESPONSE FOR DIGITAL RISKS: ARE INTERNAL AUDITORS READY?[17]

According to Dede (2016), the extensive way to initiate digital transformation in HE institutions is to influence the perspective of university staff to change. This issue determines the speed and coherence of transformation regarding the main process and sub-processes in the university; moreover, their adaptation level to the innovations. The internal audit mechanism is critical in identifying areas resistant to change arising in different departments and levels within the university and identifying potential risks to achieve an early warning system. In the context of Industry 4.0 and Education 4.0 mentioned above, it is necessary to adapt the Audit 4.0 approach to execute audit fieldworks in digitally transforming HE institutions according to the required audit standards (Dai 2017; Bozkus Kahyaoglu 2019).

It is a fact that digital transformation in universities is more than digitizing university documents. The key to success consists of joint work between the administrative and academic staff, university management, and internal auditors, reaching effective communication among all participants, including students and external stakeholders.

The coherent structure of a digitally transforming university suggests strategic forming of the digital infrastructure, learning environment, and innovative management culture. These elements are necessary for fully accomplishing the digital university framework. The increased globalization of higher education, market pressure undergoing digital transformation in university classrooms, and rapidly changing needs and expectations in e-learning, force auditors to rethink the new conditions in the audit universe and their audit culture. Hence, the auditors have additional encouragement for technology-enabled auditing (TEA) to fulfill their roles and responsibilities for providing reasonable assurance to stakeholders at the digital university.

Technology as an audit tool is not a new concept, but it has gained considerable ground in the last five to ten years. Part of the recent drive to incorporate technology in both the higher education system and audit has been a result of legislation against increasing digital risks all over the World. Data analysis, artificial intelligence, data mining, and non-linear risk assessment techniques offer much-needed efficiencies – reducing overall compliance costs and expanding the scope and reliability of audit tests. The use of these techniques also gives auditors an independent view of the higher education systems, the financial transactions of the students, administrative and academic staff, and the other key issues and stakeholders. Through continuous auditing utilizing these techniques, auditors can highlight anomalies, control deficiencies, and unusual trends. That means errors, fraud, and other problems can be identified on time.

Technological changes and innovations are transforming the critical business processes of HE institutions from teaching research and knowledge transfer. Particularly, e-learning platforms with online access to academic knowledge and learning opportunities are shifting expectations of what "bricks-and-mortar" universities will present. There are high pressures to collect first-class students from around the globe and to provide them to evolve in a digital society via digitally transforming universities (PwC 2018: 6–7). In addition, digitally transforming universities are supposed to protect excellence in research and teaching standards in an ever more competitive international marketplace. In these hard conditions, digitally transforming universities need to improve the digital capacity of their staff. They need to innovate not only the research and teaching capacity of staff but also in the pivotal business processes and in the practice of being a continuously learning organization (Jones and Goodfellow 2012: 59–60).

The main processes of HE institutions are revised to adopt digital technologies and to adjust their impacts. That leads to a higher pressure for HE institutions that are actively engaged in seeking to establish new approaches of operating that make use of the opportunities that digital technologies and innovations make available. This situation necessitates the digitally transforming HE institutions to rely on systematic performance measurement and evaluation processes based on CA/CM to achieve strategic goals (Vasarhelyi et al. 2015). This book investigates the way digital transforming universities emerge, work, and evolve, the response of the auditors to this challenging digital and riskier audit universe considering the process of digital transformation in the higher education sector.

It is a fact that there is emerging literature around digital literacies and capabilities as Sharpe and Beetham (2007); Beetham et al. (2009), Littlejohn et al. (2012). These works focus on the development of individual capabilities and digital literacies. In other words, they do not provide information about digital universities holistically at an institutional level. It is an area where rules and laws become meaningless due to the speed of digitalization. In this respect, the fundamental elements of this area will be human, and the primary directives will be ethical. However, where ethical issues exist, the conceptual audit tool is internal audit. When evaluated from this point of view, internal audit becomes the main audit tool of all digitalized processes. The main reason for this is digital operations with the potential to cause irreversible losses. Therefore, it is necessary to diagnose events in advance by diagnostic analytics and all other early warning mechanisms.

There is a need for strategic vision led by all staff with the encouragement of senior management, not just the team of Information Technology. The only tool that offers reasonable assurance in this regard is the risk-based internal audit. As a result, internal audit standards and professional practices in the world have become digital and cyber-based (Bozkus Kahyaoglu and Caliyurt 2018).

Auditors also should be familiar with risks that arise with digitalization itself. The digitally transforming university approaches should be embedded into a risk-based audit plan and Audit Committee agenda for the auditors who work in the higher education sector. The agenda needs to focus on the critical business processes and features of digital technology as an integral part of practices, tools, and techniques for the audit standards and field works (Arthur 2009). Internal auditors should focus on the persistent features of the digital to understand the digital university business model and revise their audit culture accordingly. These features include the infrastructure of the digital archive and establishment of social memory; the changeable, replicable, and malleable characteristics of digital works; the construction mechanisms of digital networks; interactivity and the interface; digital communications structure, i.e., the digital culture of participation, cooperation, and collaboration (Jones 2013).

Digitally transforming universities may have "career mismatches" and "capacity shortages" due to disturbances arising from a growing number of knowledge-based roles depending on synthesizing data and changing business models. This situation increases the operational risks, and internal auditors should consider this issue to establish early warning mechanisms. The primary aim of the auditors is to assist the formation of a framework for a more inclusive, practitioner-focused implementation for the digital transformation process in the context of the digital university. The points to be considered in the auditing of a university undergoing a digital transformation under the Committee of Sponsoring Organizations (COSO) framework are as follows (COSO 2019):

1. *Control Environment*: The business model of a digitally transforming university regarding teaching and learning, research, employment, and reward systems are changing fast. "Business models" should focus on the leading trends, relationships, and key stakeholders that influence an institution's current and future strategy and business objectives. Schlarman and Wright (2019) prepared a survey report, namely "Digital Risk Report". A key finding of the report cited that 90% of institutions in North America experiencing digital transformation accept the fact, that their risk profiles have been expanded due to their digital plans and objectives. Similarly, considering how data are shared with external parties such as outsourced service providers, increases the complexity of the audit universe, volatility of the risk universe, and dependence on an infrastructure that is not within the

Figure 1.16 Major consequences of unmanaged digital risks.

Source: RSA Digital Risk Report (2019).

supervision of the university. The consequences of digital management failures are indicated in Figure 1.16. Hence, internal auditors should be aware of both the financial and non-financial risks rising from unmanaged digital risks.

2. *Risk Assessment*: It is a fact that technologies such as the IoT, social networks, artificial intelligence (AI), and big data and analytics allow new ways to optimize performance, but they can also alter a university's risk profile. Internal auditors should note that emerging technologies and digital transformation are connected, and they provide a broader approach that better mirror the accurate profile of digital risks.
3. *Control Activities*: The university management must freely consider taking relevant risks and leverage new technologies accurately to observe opportunities available in the higher education sector. In addition, internal auditors should provide recommendations for managing the residual risks that emerge to maintain a highly secure and resilient business infrastructure. Schlarman and Wright (2019) explain the necessary steps for managing digital risks, which could be considered as control activities to achieve efficient, effective, economical, and ethical control environments (COSO 2019). These steps are summarized as

follows: (1) creating awareness about the possible digital risks, (2) analyzing digital risks, (3) discovering digital risks, (4) implementing process changes to diminish or treat the digital risks, (5) evaluating or prioritizing the digital risks utilizing relevant measurements, and (6) implementing technology changes and innovations to diminish or treat the digital risks (Balakrishnan 2017).

4. *Information and Communication*: In such a virtually connected and digitally controlled environment, the decision-making process must be made in real-time. In this context, a critical component is not only the reliability of the data but also the speed at which the data are communicated and reported. Thus, a primary digital risk may be related to an incident, which can influence the availability of an institution's all systems and underlying data that is critical for agile risk management and strategic decision-making process. Internal auditors need to have an open communication line for immediate escalation through the audit committee when a significant event or fraud occurs
5. *Monitoring*: When any change occurs in the digital platform of the university, internal auditors must consider the new digital risks that are present concerning new systems, the e-learning footprint on the digital classrooms, mobile application security, and protection of information and integrity of student registrations and all other available programs. The digital transformation process must be considered as a "business context" which evolves in the constantly changing control environment of the university. Hence, CA/CM tools and techniques are recommended for internal auditors to have value-added monitoring capabilities.

Technology is a field of swift evolution for the higher education sector since digitally transforming universities try to capture innovations and build up their strategic approaches toward technology to improve the quality of learning and teaching. Technology-enhanced learning and teaching approaches require a strong partnership with both internal and external stakeholders of digitally transforming universities. Stakeholders' experiences with technology are diversified, challenging, and innovative. In this respect, their digital expertise should be exploited as a valuable asset by digitally transforming universities. HE institutions are trying to differentiate themselves through new business models relevant to the digital business environment. That is necessary for universities that aim to become Digital Leader. It is a fact that students have become clients who request their genuine e-learning and innovation expectations from the digitally transforming universities. Thus, digitally transforming universities are required to have a space-independent structuring to translate scientific knowledge into practice and spread it digitally in society.

It is a university model in digital transformation based on compliance with ethical guidelines, instead of compliance with the law. That is because the frequency of change is high and legal compliance is lagging. The external delays in the processes experienced can lead to irreversible costs in the virtual environment. Internal auditors need to align with their university's digital transformation initiatives and strategies. Especially, they need to assess the impact of digital objectives and review relevant policies related to cybersecurity, privacy, and information governance. This approach is crucial to verify that they are still in good condition adjusting to rapidly changing business processes. Digital transformation leads to a complicated "learning curve" for internal auditors that may take time to adapt and maintain their positions in the university as trusted digital data analyzers.

NOTES

1 Välimaa (2014) defines these six revolutionary stages to explain the changes in the HE institutions in Europe. These stages are summarized as follows: the Gutenberg period (1460–1560); the Humboldt period (1810–1860); the modern research university period (1860–1920); the mass higher education period (1960–1970), and the information society period (1990–). Similarly, Wissema (2009) proposes first-generation university (1GU), second-generation university (2GU), and third-generation university (3GU) periods for the historical change pattern in the HE, respectively.

2 Neoliberalism envisages a "minimalist state" approach, and this understanding emphasizes the reduction of the regulatory and interventionist role of the state in all areas at the HE institutions (Vaira 2004).

3 Erickson and Rothberg (2014) call this relationship a "natural connection" between big data and KM because both are concerned with some intangible asset such that in the form of data, information, knowledge, or intelligence.

4 For instance, considering the KM concept for a sole entity can be defined as "personal knowledge management (PKM)", widening into "customer knowledge management (CKM)" and in the end, into "organizational knowledge management (OKM)". In this respect, it is a must for the HE sectors, and not a choice, to initiate the necessary digital transformation to ensure that the transfer of KM, PKM, CKM, and OKM to the business world is effective (Sumbal et al. 2017). In PKM, the major aim is to enhance the work efficiency of individuals, and socializing individuals by considering their personal search for gaining more information (Huang et al. 2018).

5 A social audit differs from a financial audit to the extent that the latter explores and assesses financial statements, records and their accuracy. The former, meanwhile, extends itself to qualitative aspects of service or public policy delivery; it is a broader concept. Further, social audit is usually an internally generated process since it seeks to juxtapose implementation against the stated or intended objectives of the particular institution, which can best be understood by the members of the institution itself. Social audits, by virtue of being internally operationalized, administered and assessed, bring in greater responsibility in the system as a whole, making it self-correcting and inherently transparent.

(Pareek 2014: 8–9)

Key Parameters for Social Audit and Policy Recommendations: Regularity of Class Work/ Home Work and Corrections; Bias, if any, in Student Assessment; Awareness Generation; Career Counselling: Information on Higher Studies and Training Options; Parental Involvement; Sanitation; Distribution of Syllabus Across the Year; Infrastructural Norms; Pupil-Teacher Ratio; Measures to Help Parents Assess Learning Outcomes; Environment-Building; Extra-Curricular Activities; Age-Appropriate Admissions and Student Absenteeism.

(Pareek 2014: 37)

6 Sponsorship of digital transformation initiatives is migrating to the top of organizations and laterally from CIO to CMO, COO, and even CEO. At the same time, CAEs and internal auditors are fighting with how to keep their risk-based auditing strategies relevant amid the rapid step of change caused by the digital transformation.

7 Fisk (2017) introduces "Education 4.0" as a vision for the future of education based on the following assumptions: Education 4.0 is defined as a response

- To the needs of Industry 4.0, where humans and machines align to work based on a collaborative approach
- Utilize the benefits of personalized data and open-source data
- Set up a design for the future of learning – including lifelong learning from childhood to continuous learning through the workplace
- Achieve a better teaching and learning role in society

8 It should be noted that these e-business classifications delineate the digital business activities considering their "ideal forms". In other words, the division of e-commerce, e-collaboration, e-communication, e-education, and e-information/entertainment reflects a general framework of digital business from a theoretical and conceptual perspective. However, organizations usually apply these digital business activities in combination with their corporate practice. In this respect, it is difficult to make a clear distinction between them since overlaps may take place.

9 International Association of Universities can be defined as follows:

> IAU is a membership-led organization to advance higher education and its important role in the development of our society. Founded in 1950 under the auspices of UNESCO, it is the leading global association of higher education institutions and organizations, comprising more than 650 Members in some 130 countries. IAU serves as a global forum for leaders in higher education to reflect and rally around common priorities. It acts as the voice of higher education to UNESCO and other international organizations.

10 Evolutionary innovation is defined as "continuous or dynamic evolutionary innovation" which is introduced by incremental advances in technology or processes.

11 Revolutionary innovation is defined as "discontinuous innovation" and for this reason, it is usually seen as "disruptive and new".

12 Academic learning is used for assisting institutional operations and strategic decision-making processes.

13 Educational data mining is used for the development and evaluation of data analytics methods for investigating educational data.

14 Learning analytics is used for the implementation of data analytic tools and techniques to examine and enhance learning and teaching.

15 CERN Operating with a budget comparable to a medium-sized European university, CERN is one of the world's leading centers for particle physics and produces cutting-edge science and technology. The unique know-how and expertise of CERN scientists and engineers provide excellent opportunities to bridge the gap between science and society. This happens in many ways, one of which is by harnessing technological advances to create concrete solutions for industry in many fields from medical and biomedical technologies to aerospace applications, and from Industry 4.0 to cultural heritage (please visit the CERN website for more information).

16 The definition by Wikipedia for Massive Open Online Course (MOOC) is

> an online course aimed at unlimited participation and open access via the web. In addition to traditional course materials such as videos, readings, and problem sets, MOOCs provide interactive user forum? that help builds a community for students, professors, and teaching assistants

For further details, please visit Wikipedia.org.

17 The main topic of this section was presented as a paper, namely "*Rethinking Audit Culture In Digital Transformation Era: Recommendations For Auditing Digitally Transforming Universities*" by the Authors in the 11th International Conference on Governance Fraud Ethics and Corporate Social Responsibility (CSR) which was virtually hosted by London Metropolitan University, Guildhall School of Business and Law on 27–28 August 2020.

REFERENCES

Abel, J. R., Deitz, R. (2010). The role of colleges and universities in building local human capital. Federal Reserve Bank of New York. *Current Issues in Economics and Finance*, 17(6), 1–7.

Amit, R., Zott, C. (2012). Creating value through business model innovation. *MIT Sloan Management Review*, 53(3), 40–50.

Andreeva, T., Kianto, A. (2012). Does knowledge management really matter? Linking knowledge management practices, competitiveness and economic performance. *Journal of Knowledge Management*, 16(4), 617–636.

Arthur, W. B. (2009). *The nature of technology: What it is and how it evolves*. London: Allen Lane.

Avella, J. T., Kebritchi, M., Nunn, S. G., Kanai, T. (2016). Learning analytics methods, benefits, and challenges in higher education: A systematic literature review. *Online Learning*, 20(2), 13–29.

Baden-Fuller, C., Morgan, M. S. (2010). Business models as models. *Long Range Planning*, 43 (2–3), 156–171.

Baker, R. (2010). Data mining for education. *International Encyclopedia of Education*, 7, 112–118.

Baker, R., Inventado, P. S. (2014). Educational data mining and learning analytics. In J. A. Larusson & B. White (Eds.), *Learning analytics: From research to practice* (pp. 61–75). New York, NY: Springer.

Balakrishnan R. (2017). Internal audit in the age of digital transformation. *Potiviti.* https://chapters.theiia.org/IIA%20Canada/Thought%20Leadership%20Documents/Internal-Audit-in-the-Age-of-Digital-Transformation.pdf
van Barneveld, A., Arnold, K., Campbell, J. P. (2012). Analytics in higher education: Establishing a common language. *EDUCAUSE*. Retrieved from https://library.educause.edu/resources/2012/1/analyticsin-higher-education-establishing-a-common-language
Bates, T. (2010). New challenges for universities: Why they must change? In U. D. Ehlers & D. Schneckenberg (Eds.), *Changing cultures in education moving ahead to future learning* (pp. 15–25). Heidelberg: Springer-Verlag.
Beetham, H., McGill, L., Littlejohn, A. (2009). *Thriving in the 21st century: Learning literacies for the digital age (LLiDA Project) executive summary, conclusions and recommendations*. UK Joint Information Systems Committee (JISC). http://oro.open.ac.uk/52237/
Beheshti, H. M., Salehi-Sangari, E. (2007). The benefits of e-business adoption. An empirical study of Swedish SMEs. *Service Business*, 1(3), 233–245.
Ben-David, J., Zloczower, A. (1962). Universities and academic systems in modern societies. *European Journal of Sociology/Archives Européennes de Sociologie*, 3(1), 45–84.
Bernhard, A. (2012). Quality Assurance in an International Higher Education Area. Dissertation University of Klagenfurt, Springer Fachmedien Wiesbaden GmbH. ISBN:978-3-531-18566-8.
Bizri, R., Hammoud, J., Stouhi, M., Hammoud, M. (2019). The entrepreneurial university: A proposed model for developing nations. *Journal of Management Development*, 38(5), 383–404.
Bleiklie, I., Kogan, M. (2006). Comparison and theories. In M. Kogan, M. Bauer, I. Bleiklie, & M. Henkel (Eds.), *Transforming higher education: A comparative study* (2nd ed., pp. 3–22). Dordrecht: Springer.
Bozkus Kahyaoglu, S. (2019). An analysis on the implementation of new approaches and techniques in the auditing of business processes based on Blockchain technologies. In B. Darici & F. Ayhan (Eds.), *Cryptocurrencies in all aspects*. New York: Peterlang Publications. ISBN:978-3-631-78387-0.
Bozkus Kahyaoglu, S., Caliyurt, K. (2018). Cyber security assurance process from the internal audit perspective. *Managerial Auditing Journal*, 33(4), 360–376.
Bozkus Kahyaoglu, S., Sarıkaya, R., Topal, B. (2020). Continuous auditing as a strategic tool in public sector internal audit: The Turkish case. *Journal of Selçuk University Social Sciences Vocational School* (Selçuk Üniversitesi Sosyal Bilimler Meslek Yüksekokulu Dergisi), 23(1), 208–225. DOI: 10.29249/selcuksbmyd.670261
Bradley, J., Loucks, J., Macaulay, J., Noronha, A., Wade, M. (2015). *Digital vortex: How digital disruption is redefining industries*. Global Center for Digital Business Transformation: An IMD and Cisco Initiative. June.
Buckingham Shum, S., Ferguson, R. (2012). Social learning analytics. *Educational Technology & Society*, 15(3), 3–26.
Campbell, D. F. J., Carayannis, E. G. (2013). *Epistemic governance in higher education quality enhancement of universities for development*. New York: Springer Brief in Business Publications. ISBN:978-1-4614-4417-6.
Cantú-Ortiz, F. J. (2018). *Research analytics boosting university productivity and competitiveness through scientometrics*. Boca Raton, FL: Taylor & Francis Group, LLC. ISBN:978-1-4987-8542-6.

CERN – European Council for Nuclear Research (CERN) (2017). *Knowledge transfer*. Annual report. https://home.cern/resources/annual-report/knowledge-sharing/knowledge-transfer-annual-reports

Cevik Onar, S., Ustundag, A., Kadaifci, Ç., Oztaysi, B. (2018) The changing role of engineering education in Industry 4.0 Era. In *Industry 4.0: Managing the digital transformation*. Springer Series in Advanced Manufacturing. Cham: Springer. DOI: 10.1007/978-3-319-57870-5_8

Chaurasia, S. S., Kodwani, D., Lachhwani, H., Ketkar, M. A. (2018). Big data academic and learning analytics. *International Journal of Educational Management*, 32(6), 1099–1117.

Cleary, J., Van Noy, M. (2014). A framework for higher education labor market alignment: Lessons and future directions in the development of jobs-driven strategies. Working Paper, New Brunswick Heldrich Center for Workforce Development, Rutgers University, New Brunswick, New Jersey.

Coaldrake, P., Stedman, L. (2016). *Raising the stakes: Gambling with the future of universities* (2nd ed.). Brisbane: University of Queensland Press.

COSO. (2019). Managing cyber risk in a digital age. https://www.coso.org/Documents/COSO-Deloitte-Managing-Cyber-Risk-in-a-Digital-Age.pdf

Crane, L., Self, R. J. (2014). Big data analytics: A threat or an opportunity for knowledge management? In L. Uden et al. (Eds.), *International conference on knowledge management in organizations* (pp. 25–34), September. Cham: Springer.

Dahlstrom, E., Brooks, D. C., Bichsel, J. (2014). The current ecosystem of learning management systems in higher education: Student, faculty, and IT perspectives. *EDUCAUSE Review*. Retrieved from http://www.educause.edu/ecar.2014

Dai, J. (2017). Three essays on audit technology: Audit 4.0, blockchain, and Audit App. PhD Dissertation. The State University of New Jersey, Rutgers. Retrieved from https://rucore.libraries.rutgers.edu/rutgers-lib/55154/

Daniel, B. (2015). Big data and analytics in higher education: Opportunities and challenges. *British Journal of Educational Technology*, 46 (5), 904–920.

Daniel, B. K., Butson, R. (2013). Technology enhanced analytics (TEA) in higher education. In *Proceedings of the international conference on educational technologies* (pp. 89–96), 29 November–1 December, 2013, Kuala Lumpur, Malaysia.

Dede, C. (2016). Designing an educational innovation to achieve scale: Five critical concepts. The evolution: A destiny solutions illumination. https://evolllution.com/managing-institution/operations_efficiency/designing-an-educational-innovation-to-achieve-scale-five-critical-concepts/

Deem, R. (2001). Globalisation, new managerialism, academic capitalism and entrepreneurialism in universities: is the local dimension still important? *Comparative Education*, 37(1), 7–20. https://www.jstor.org/stable/3099730 (Acccss date: 20200729).

Denger, K., Wirtz, B. W. (1995). Innovatives wissensmanagement und multimedia. *Gablers Magazin*, 9(3), 20–24.

DeWit, B., Meyer, R. (2010). *Strategy synthesis—Resolving strategy paradoxes to create competitive advantage* (2nd ed.). Hampshire: South-Western Cengage Learning.

DFWG (Digital Futures Working Group) (2014). *Digital futures working group: Recommendations*, April 2014 (Final Revision). Edinburgh Napier University.

Dringus, L. P. (2012). Learning analytics considered harmful. *Journal of Asynchronous Learning Networks*, 16(3), 87–100.

Dyckhoff, A. L., Zielke, D., Bültmann, M., Chatti, M. A., Schroeder, U. (2012). Design and implementation of a learning analytics toolkit for teachers. *Educational Technology & Society*, 15(3), 58–76.

Erickson, S., Rothberg, H. (2014). Big data and knowledge management: Establishing a conceptual foundation. *Electronic Journal of Knowledge Management*, 12(2), 101.

Faviero B. B. F., (2012). *Major players in online education market. Comparing Khan Academy, Coursera, Udacity, & edX missions, offerings*. The Tech, Vol. 132, no. 34, September 2012.

Fisk, P. (2017). Education 4.0. http://www.thegeniusworks.com/2017/01/future-education-youngeveryone-taught-together/

George, E. S. (2006). Positioning higher education for the knowledge-based economy. *Higher Education*, 52(4), 589–610.

Greller, W., Drachsler, H. (2012). Translating learning into numbers: A generic framework for learning analytics. *Educational Technology & Society*, 15(3), 42–57.

Hemsley-Brown, J. (2011). Market, heal thyself: The challenges of a free market in higher education. *Journal of Marketing for Higher Education*, 21(2), 115–132.

Horta, H. (2009). Global and national prominent universities: internationalization, competitiveness and the role of the state. *Higher Education*, 58, 387–405.

Huang, Y. M., Pauleen, D., Scahill, S., Taskin, N. (2018). A PKM-based decision-making training program for personal healthcare: An action learning approach. *International Journal of Knowledge Management*, 14(3), 101–114.

Hulla, M., Karre, H., Hammer M., Christian, R. (2019). A teaching concept towards digitalization at the LEAD factory of Graz University of Technology. In M. E. Auer & T. Tsiatsos (Eds.), *The challenges of the digital transformation in education Proceedings of the 21st International Conference on Interactive Collaborative Learning (ICL2018)* – Vol. 2, AISC 917 (pp. 393–402), 2019. DOI: 10.1007/978-3-030-11935-5_38

Ifenthaler, D. (2015). Learning analytics. In J. Spector (Ed.), *The SAGE Encyclopedia of Educational Technology* (pp. 448–451). Thousand Oaks, CA: SAGE Publications, Inc.

Jensen, T. (2019). Higher education in the digital era. The current state of transformation around the world in the digital era. International Association of Universities Survey Report. ISBN:978-92-9002-210-7.

Jeschke, S., Heinze, U. (2014). Higher Education 4.0 – Trends and future perspectives for teaching and learning. DOI: 10.13140/RG.2.1.1509.0002. Conference: 64. DHV-Tag Virtuelle Lernwelten in der UniversitätAt: Frankfurt am Main. https://www.researchgate.net/publication/284717997_Higher_Education_40_-_Trends_and_Future_Perspectives_for_Teaching_and_Learning

Jessop, B. (2017). Varieties of academic capitalism and entrepreneurial universities on past research and three thought experiments. *Higher Education*, 73, 853–870. https://link.springer.com/content/pdf/10.1007/s10734-017-0120-6.pdf (Access date: 20200831).

Jones, C. (2013). The digital university: A concept in need of a definition. In R. Goodfellow & M. Lea (Eds.), *Literacy in the digital university—Critical perspectives on learning, scholarship and technology* (pp. 162–172). London: Routledge.

Jones, C., Goodfellow, R. (2012). The "Digital University": Discourse, theory and evidence. *International Journal of Learning and Media*, 4(3–4), 60. https://www.researchgate.net/publication/275504610_The_Digital_University_Discourse_Theory_and_Evidence

Jones, S. (2012). Technology review: The possibilities of learning analytics to improve learner-centered decision-making. *Community College Enterprise*, 18(1), 89–92.

Kahneman, K. (2011). *Thinking, fast and slow*. USA: Penguin Group. ISBN:978-1-846-14055-6.

Kiss, V., Vass, V. (2019). The transformative role of innovation in the higher education. In M. E. Auer & T. Tsiatsos (Eds.), *The challenges of the digital transformation in education proceedings of the 21st International Conference on Interactive Collaborative Learning (ICL2018)* – Vol. 2. Springer. ISBN:978-3-030-11934-8.

Kulkarni, C., Wei, K.P., Le, H., Chia, D., Papadopoulos, K., Cheng, J., Koller, D., Klemmer, S. R. (2015). Peer and self-assessment in massive online classes. In: *Design thinking research* (pp. 131–168). Switzerland: Springer International Publishing.

Kumaran, D., Hassabis, D., McClelland, J. L. (2016). What learning systems do intelligent agents need? Complementary learning systems theory. *Updated in Trends in Cognitive Sciences*, 20(7), 512–534.

Little, D. A. (2016). The future of higher education. Transforming the students of tomorrow. https://www.adlittle.com/Thefutureofhighereducation

Littlejohn, A., Beetham, H., McGill, L. (2012). Learning at the digital frontier: A review of digital literacies in theory and practice. *Journal of Computer Assisted Learning*, 28(6), 547–556.

MacNeill, S., Johnston, B. (2012). A conversation around what it means to be a digital university. http://blogs.cetis.org.uk/sheilamacneill/2012/01/26/a-converstaion-around-what-it-means-to-be-a-digital-university/

Malseva, A. (2018). Intellectual capital of universities and regions: A qualitative analysis of mutual influence's directions. *Education*, 39(20), 18–32. https://www.revistaespacios.com/a18v39n20/a18v39n20p18.pdf

Manyika, J., Chui, M., Brown, B., Bughin, J., Dobbs, R., Roxburgh, C., Byers, A. H. (2011). *Big data: The next frontier for innovation, competition, and productivity*. USA: McKinsey Global Institute. http://www.mckinsey.com/Insights/MGI/Research/Technology_and_Innovation/Big_data_The_next_frontier_for_innovation

Marginson, S. (2013). The impossibility of capitalist markets in higher education. *Journal of Education Policy*, 28(3), 353–370.

Martin, M., Parikh, S. (2017). Quality management in higher education: Developments and drivers. *Results from an international survey*. Paris, France: UNESCO Publication. ISBN:978-92-803-1412-0.

Milliken, J., Colohan, G. (2004). Quality or control? Management in higher education. *Journal of Higher Education Policy and Management*, 26(3), 1–37.

Nguyen, A., Gardner, L., Sheridan, D. (2017). A multi-layered taxonomy of learning analytics applications. In *Proceedings of the Pacific Asia conference on information systems*, Article 54.

Nguyen, A., Gardner, L., Sheridan, D. (2020). Data analytics in higher education: An integrated view. *Journal of Information Systems Education*, 31(1) Winter 2020, 61–71.

Nissen, V. (2018). *Digital transformation of the consulting industry extending the traditional delivery model*. Switzerland: Springer. ISBN:978-3-319-70490-6.

Nistor, N. & Hernández-Garcíac, Á. (2018). What types of data are used in learning analytics? An overview of six cases. *Computers in Human Behavior*, 89, 335–338.

Nybom, T. (2007). A rule-governed community of scholars: The Humboldt vision in the history of the European university. In P. Maassen & J. P. Olsen (Eds.), *University dynamics and European integration* (pp. 55–79). Dordrecht: Springer.

OECD. (2002). Frascati Manual 2002. The measurement of scientific and technological activities. Proposed standard practice for surveys on research and experimental development. OECD Publication.

OECD. (2012). *Better skills, better jobs, better lives: A strategic approach to skills policies*. Paris, France: OECD Publishing. http://www.oecd.org/education/imhe/IMHEinfos_Jult12_EN%20-%20web.pdf

OECD. (2013). New sources of growth: Knowledge-based capital – Key analyses and policy conclusions – Synthesis report. https://www.oecd.org/sti/inno/knowledge-based-capital-synthesis.pdf

OECD. (2020). Education at a glance. https://read.oecd-ilibrary.org/education/education-at-a-glance-2020_69096873-en#page5

Pareek, V. (2014). Social audit framework for the education sector. Researching Reality Summer Internship 2014 Working paper: 330. https://ccsinternship.files.wordpress.com/2014/06/330_social-audit-framework-for-the-education-sector_vrinda-pareek.pdf

Pauleen, D. J., Wang, W. Y. (2017). Does big data mean big knowledge? KM perspectives on big data and analytics. *Journal of Knowledge Management*, 21(1), 1–6.

Pee, L. G., Kankanhalli, A. (2009). A model of organisational knowledge management maturity based on people, process, and technology. *Journal of Information & Knowledge Management*, 8(2), 79–99.

Peña-Ayala, A. (2014). Educational data mining: A survey and a data mining-based analysis of recent works. *Expert Systems with Applications*, 41(4), 1432–1462.

Picciano, A. G. (2014). Big data and learning analytics in blended learning environments: Benefits and concerns. *International Journal of Artificial Intelligence and Interactive Multimedia*, 2(7), 35–43.

Pomeroy, W. L. (2014). Academic analytics in higher education: Barriers to adoption. PhD Thesis. Walden University, Minneapolis, MN.

PwC. (2018). *The 2018 digital university staying relevant in the digital age*. UK: PWC. https://www.pwc.co.uk/assets/pdf/the-2018-digital-university-staying-relevant-in-the-digital-age.pdf

Riddell, A. R. (1996). Globalization: emasculation or opportunity for educational planning? *World Development*, 24(8), 1357–1372.

Romero, C. R., Ventura, S. (2010). Educational data mining: a review of the state of the art. *IEEE Transactions on Systems, Man and Cybernetics, Part C: Applications and Reviews*, 40(6), 601–618.

Rouhiainen, L. (2019). How AI and data could personalize higher education. *Harvard Business Review*. https://hbr.org/2019/10/how-ai-and-data-could-personalize-higher-education

Sala-i-Martín, X., Crotti, R., di Battista, A., Hanouz, M. D., Galvan, C., Geiger, T., Marti, G. (2015). Reaching beyond the new normal: Findings from the global competitiveness index 2015–2016. In K. Schwab & X. Sala-iMartín (Eds.), *The global competitiveness report 2015–2016*. Geneva: World Economic Forum.

Salem, M. I. (2014). The role of universities in building a knowledge-based economy in Saudi Arabia. *International Business & Economics Research Journal*, 13(5), 1047–1056. DOI: 10.19030/iber.v13i5.8771

Salmi, J. (2009). *The challenge of establishing world-class universities.* Washington, DC: World Bank.
Sambamurthy, V., Zmud, R. W. (2017). *Guiding the Digital Transformation of Organizations.* 2nd Edition, Legerity Digital Press. ISBN:978-0-9995347-0-0.
Sánchez-Barrioluengo, M. (2013). Articulating the 'Three-Missions' in Spanish universities. *Research Policy*, 43 (10), 1760–1773.
Schlarman, S., Wright, J. (2019). *Digital risk report*, 1st ed. RSA. https://www.rsa.com/content/dam/en/white-paper/rsa-digital-risk-report-2019.pdf
Schumpeter, J. (1939). *Business cycles: A theoretical, historical and statistical analysis of the capitalist process.* New York and London: McGraw – Hill Book Company Inc. ISBN:978-1-68422-065-6.
Sharpe, R., Beetham, H. (Eds.). (2007). *Rethinking pedagogy for a digital age.* Hoboken: Taylor and Francis.
Siemens G. (2012). Connectivism: A learning theory for the digital age, Jan-2005. [Online]. Available: http://www.itdl.org/Journal/Jan_05/article01.htm
Siemens, G. (2013). Learning analytics: The emergence of a discipline. *American Behavioral Scientist*, 57(10), 1380–1400.
Siemens, G., Long, P. (2011). Penetrating the fog: Analytics in learning and education. *EDUCAUSE Review*, 46(5), 30.
Sin, K., Muthu, L. (2015). Application of big data in education data mining and learning analytics – A literature review. *ICTACT Journal on Soft Computing*, 5(4), 1035–1049.
Slade, S., Prinsloo, P. (2013). Learning analytics: Ethical issues and dilemmas. *American Behavioral Scientist*, 57(10), 1509–1528.
Slaughter, S., Leslie, L. L. (1997). *Academic capitalism: Politics, policies, and the entrepreneurial university.* Baltimore, MD: Johns Hopkins University Press.
Slaughter, S., Rhoades, G. (2004). *Academic capitalism and the new economy: Markets, state, and higher education.* Baltimore, MD: Johns Hopkins University Press.
Slavin, R. E. (1990). *Cooperative learning: Theory, research, and practice.* Hoboken, NJ: Prentice-Hall.
Smyth, K., MacNeill, S., Johnston, B. (2015). Visioning the digital university – From institutional strategy to academic practice. *Educational Developments*, 16(2), 13–17.
Stone, S. M. (2019). *Digitally deaf, why organizations struggle with digital transformation.* Switzerland: Springer. ISBN:978-3-030-01832-0.
Sumbal, M. S., Tsui, E., See-to, E. W. (2017). Interrelationship between big data and knowledge management: An exploratory study in the oil and gas sector. *Journal of Knowledge Management*, 21(1), 180–196.
Tapper, T., Palfreyman, D. (2004). *Convergence and divergence in the global model of mass higher education: Predictions for 2010.* Oxford: CHEPS.
Thomas, A., Chopra, M. (2020). On how big data revolutionizes knowledge management (Chapter 3). In B. George & J. Paul (Eds.), *Digital transformation in business and society theory and cases.* Palgrave Macmillan. ISBN:978-3-030-08276-5.
Trequattrini, R., Lombardi, R., Lardo, A., Cuozzo, B. (2015).The impact of entrepreneurial universities on regional growth: A local intellectual capital perspective. *Journal of the Knowledge Economy*, 1–13.
Turban, E., King, D., Lee, J. K., Liang, T. -P., Turban, D. C. (2015). *Electronic commerce: A managerial and social networks perspective.* Berlin: Springer.

Vaira, M. (2004). Globalization and higher education organizational change: A framework for analysis. *Higher Education*, 48(4), 483–510.

Välimaa, J. (2014). University revolutions and academic capitalism a historical perspective. In B. Cantwell & I. Kauppinen (Eds.), *Academic capitalism in the age of globalization* (pp. 31–51). Baltimore, MD: Johns Hopkins University Press.

Vasarhelyi, M. A. (2002). Concepts in continuous assurance. In S. Sutton & V. Arnold (Eds.), *Researching accounting as an information systems discipline*. Sarasota, FL: American Accounting Association.

Vasarhelyi, M. A., Alles, M. G., Kuenkaikaew, S., Littley, J. (2012). The acceptance and adoption of continuous auditing by internal auditors: A micro analysis. *International Journal of Accounting Information Systems*, 13(3), 267–281.

Vasarhelyi, M. A., Halper, F. B. (1991). The continuous audit of online systems. *A Journal of Practice & Theory*, 10(1):110–125. https://www.researchgate.net/profile/Miklos_Vasarhelyi/publication/255667612_The_Continuous_Audit_of_Online_Systems/links/5410341a0cf2f2b29a3f4fbf/The-Continuous-Audit-of-OnlineSystems.pdf?origin=publication_detail

Vasarhelyi, M. A., Kuenkaikaew, S. (2010). *Continuous auditing and continuous control monitoring: Case studies from leading organizations*. Rutgers Business School, Rutgers Accounting Research Center.

Vasarhelyi, M. A., Michael, G. A., Alexander, K. (2004). Principles of analytic monitoring for continuous assurance. *Journal of Emerging Technologies in Accounting*, 1(1), 1–21.

Vasarhelyi, M. A., et al. (2015). Big data in accounting: An overview. *Accounting Horizons*, 29(2): 381–396.

Westerheijden, F. D., Kohoutek, J. (2014). Implementation and translation: From European Standards and guidelines for quality assurance to education quality work in higher education institutions. In H. Eggins (Ed.), *Drivers and barriers to achieving quality in higher education*. Sense Publishers. ISBN:978-94-6209-494-9.

Wirtz, B. W. (2018). *Business model management: Design—Instrumente – Erfolgsfaktoren* (4th ed.). Wiesbaden: Springer Gabler.

Wirtz, B. W. (2019). *Digital business models concepts, models, and the alphabet case study*. Springer Publications, ISBN:978-3-030-13004-6.

Wirtz, B. W., Daiser, P. (2017). Business model innovation: An integrative conceptual framework. *Journal of Business Models* 5(1), 14–34.

Wirtz, B. W., Göttel, V., Daiser, P. (2016). Business model innovation: Development, concept, and future research directions. *Journal of Business Models*, 4(1), 1–28.

Wissema, J. G. (2009). *Towards the third generation university managing the university in transition*. Edward Elgar Publications. ISBN:978-1-84844-216-0.

World Bank. (2011). *Learning for all: Investing in people's knowledge and skills to promote development*. Washington, DC: World Bank.

World Bank. (2012). The four pillars of a knowledge economy. *Knowledge Assessment Methodology*, 37(June 21), 49–76.

World Bank. (2016). *World Bank Group industry-specific support to promote industrial competitiveness and its implications for jobs*. Washington, DC: World Bank.

World Bank Independent Evaluation Group (IEG) (2017). *Higher education for development an evaluation of the World Bank Group's support an independent evaluation*. World Bank Publications. https://openknowledge.worldbank.org/handle/10986/26486

Yang, D.-H., You, Y.-Y., Kwon, H.-J. (2014). A framework for business model innovation using market, component and innovation tool. *International Journal of Applied Engineering Research*, 9(21), 9235–9248.

Zouaq, A., Joksimovic, S., Gasevic, D. (2013). Ontology learning to analyze research trends in learning analytics publications. *LAK (Data Challenge) conference*.

Chapter 2

Visioning the digital era from the audit perspective

2.1 LOCATING "THE DIGITAL" IN A COMPETITIVE ENVIRONMENT

It is possible to express the business model as a fundamental tool to create value and maximize profitability incorporate functioning (Hess et al. 2016). Hinssen (2010) argues that organizations should reinvent their business models and locate themselves as a "platform" in the "new normal" ecosystems with the current digital trends. When an organization transforms its business model to become a platform, it also needs to transform its relationship with the stakeholders. And it needs to change its approach for product- and service-delivery processes as well. Because platform setup changes the value-generation process for the organizations in the new normal ecosystems. Platform setup extends the value-generation process to increase the active involvements of all actors by assigning them roles as both producers and consumers which is called "prosumers"[1] in the ecosystems simultaneously. In this context, it is important that organizations become more "open". However, "openness" can also mean a security issue for organizations. To prevent such a security vulnerability, organizations must develop their IT infrastructure and perform digital transformation as a business strategy (Stolterman and Fors 2004; Messina 2018).

Tremblay et al. (2012) state that the technology layer is proposed to establish new value-generation processes by using open Application Programming Interfaces (APIs)[2] which can provide easy interactions and authorizations in the platforms among all the actors. It should be noted that the meaning of the technology layer here is different from the conventional way of business.

The new meaning of technology layer has a strong position with full integration of all IT capabilities and more effective tool for competing in the new normal ecosystems rather than the past concept of technology layer which used to mean as a part of business process automation (Uhl and Gollenia 2014; Capitani 2018).

DOI: 10.1201/9781003093008-2

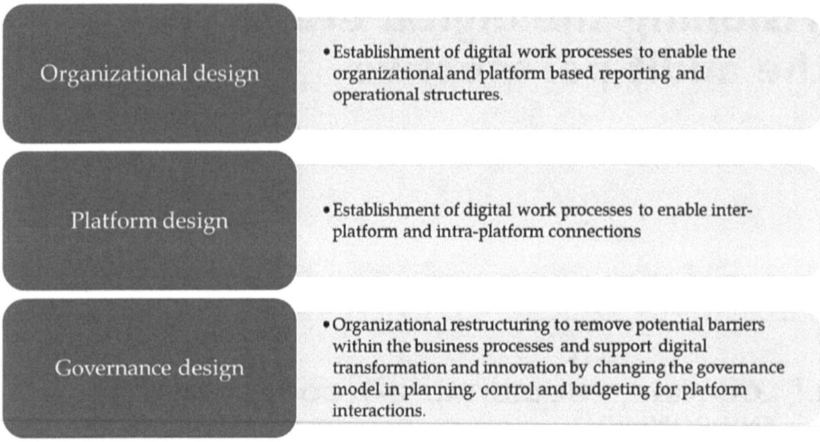

Figure 2.1 The design elements of digital transformation.

Source: Sambamurthy and Zmud (2017) and Brooks and McCormack (2020).

With the use of APIs, asset capabilities will be virtualized, applications or service implementations will be separated from their end-users, and it will provide a powerful tool for third parties to create additional services (Sambamurthy and Zmud 2017). In this respect, digital transformation should take place with an integrated approach to achieve strategic goals in a competitive environment. Sambamurthy and Zmud (2017) present a proposal to reveal the "organizational design", "platform design", and "governance design" perspectives within an integrated approach which is summarized in Figure 2.1. In this context, Pandey and Kumar (2020) call the choice of technology a "game-changer" for the HE sectors.

Digital transformation is defined by Brooks and McCormack (2020) as "deep technological shifts that enable new teaching, learning and operating models, integrated with the culture. And this transforms an institution's strategic directions, activities, and value statement". It should be noted that the concept of "digitization[3]", "digitalization[4]", and "digital transformation[5]" have different meanings. These concepts can be used for determining the "digital maturity level" of the organization (Stolterman and Fors 2004).

It is not easy to carry out all these digital transformation processes in an integrated and effective manner. For this, it is important to adapt the "organization culture" to the "digitalization culture" (Bührig et al. 2018; Brooks and McCormack 2020). McLachlan et al. (2017) present their approach for the digital culture in the university as follows:

- First, each and every educator will be a digital educator
- Second, students and all learners will be digital

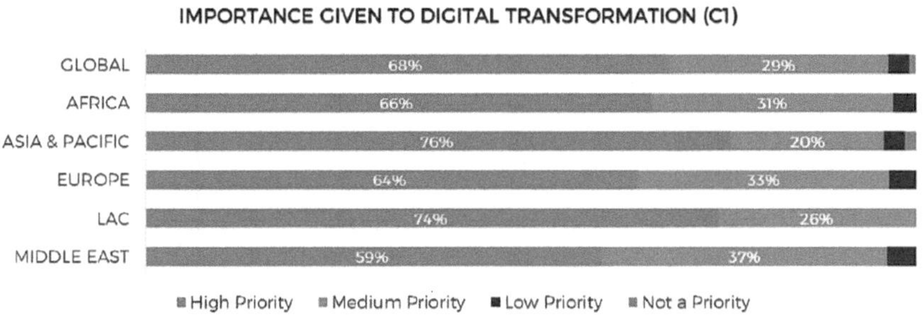

Figure 2.2 Commitment to digital transformation in the HE sectors.

Source: IAU (2019).

- Third, each and every person will improve and update their digital skills. In this way, there will be no need to worry about the future and universities will be leading their own digitalization process by making accurate future predictions

International Association of Universities (IAU) (2019) has recently prepared a research report to evaluate the effects of digital transformation on the commitment, risk, and culture of HE institutions all over the world. Even if the response rate of countries participating in the survey differs, the average level of management support is reasonably high as shown in Figure 2.2. Accordingly, leaderships in higher education institutions around the world confirm the importance of digitalization and the necessity of transformation (IAU 2019).

Although the level of management and leadership support, and their commitment is evident in the empirical findings of the IAU Research Report (2019), it is revealed that there are different approaches to how digital transformation should be translated into action in their HE institutions. These approaches are categorized into two groups. The first one is the "top-down approach" which is pushed by the leadership in the university, and the second one is based on the "bottom-up approach" with the participation of all employees throughout the university (Figure 2.3).

It is seen from the Research Report of IAU (2019) that the digitalization of HE institutions varies as a result of two different implementation processes. One of these practices is implemented by the direction of a particular leadership or the state by a "top-down approach". This situation is oriented toward the Continental European system, given its economic developments. The desired goal of this approach cannot be explained in terms of a cost–benefit analysis.

The second form of application is based on the "bottom-up approach". In this approach, excessive production is realized leading to a better status which is achieved by combining human relations and production in an

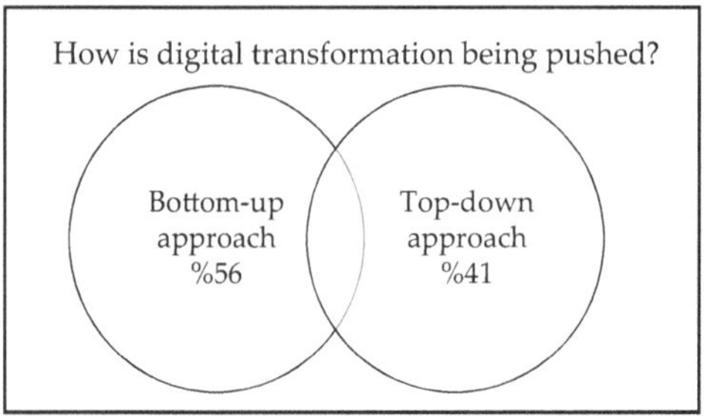

Figure 2.3 Digital transformation approach in the HE sectors.

Source: IAU (2019).

integrated manner. In addition, this approach is independent and involves more capabilities that come to the fore during the application processes. In this process, it is seen that when acting independently and having a horizontal organizational structure are combined with internal and external dynamics, then the dynamism will be provided by implicit knowledge (Apraxine and Stylianou 2017). In this way, there will be a structure that accelerates the process automatically emerges.

This process has created a new kind of platform economies that commercialize and use more knowledge for universities or the system of economic agents that are composed of a network of relations within the ecosystem and turned into a mechanism that accelerates the functioning of this system. In this process, the knowledge is transferred between university faculty members and companies which has an effect leading the universities to evolve and change (Rajterič 2010). Examples of scientists who work as academics and represent this new work environment and processes in the HE institutions are as follows:

- William T. Freeman – Professor of Electrical Engineering and Computer Science (EECS) at MIT, and a member of the Computer Science and Artificial Intelligence Laboratory (CSAIL), Research manager in Google Research in Cambridge, MA. He received outstanding paper awards at computer vision or machine learning conferences in 1997, 2006, 2009, 2012, and 2019, and test-of-time awards for papers in 1990, 1995, and 2005. He shared the 2020 Breakthrough Prize in Physics for a consulting role with the Event Horizon Telescope collaboration, which reconstructed the first image of a black hole. He is a Fellow of IEEE, ACM, and AAAI. In 2019, he received the PAMI Distinguished Researcher Award, the highest award in computer vision. He holds over 50 patents. https://billf.mit.edu/about/bio
- Alex "Sandy" Pentland – Toshiba Professor of Media Arts & Science and Professor of Information Technology. Alex "Sandy" Pentland

directs MIT's Human Dynamics Laboratory and the MIT Media Lab Entrepreneurship Program, co-leads the World Economic Forum Big Data and Personal Data initiatives, and is a founding member of the Advisory Boards for Nissan, Motorola Mobility, Telefonica, and a variety of start-up firms. He has previously helped create and direct MIT's Media Laboratory, the Media Lab Asia laboratories at the Indian Institutes of Technology, and Strong Hospital's Center for Future Health. In 2012, Forbes named Sandy one of the "seven most powerful data scientists in the world", along with Google founders and the CTO of the United States, and in 2013, he won the McKinsey Award from Harvard Business Review. He is among the most-cited computational scientists in the world, and a pioneer in computational social science, organizational engineering, wearable computing (Google Glass), image understanding, and modern biometrics. His research has been featured in Nature, Science, and Harvard Business Review, as well as being the focus of TV features on BBC World, Discovery and Science channels. Sandy's research group and entrepreneurship program have spun off more than 30 companies to date, three of which are publicly listed and several that serve millions of poor in Africa and South Asia. https://mitsloan.mit.edu/faculty/directory/sandy-pentland

- Kristen Grauman – Professor in the Department of Computer Science at the University of Texas at Austin and a Research Scientist in Facebook AI Research (FAIR). Her research in computer vision and machine learning focuses on visual recognition and search. Before joining UT-Austin in 2007, she received her Ph.D. from MIT. She is an IEEE Fellow, AAAI Fellow, Sloan Fellow, a Microsoft Research New Faculty Fellow, and a recipient of NSF CAREER and ONR Young Investigator awards, the PAMI Young Researcher Award in 2013, the 2013 Computers and Thought Award from the International Joint Conference on Artificial Intelligence (IJCAI), the Presidential Early Career Award for Scientists and Engineers (PECASE) in 2013. She was inducted into the UT Academy of Distinguished Teachers in 2017. She and her collaborators have been recognized with several Best Paper awards in computer vision, including a 2011 Marr Prize and a 2017 Helmholtz Prize (test of time award). She currently serves as an Associate Editor-in-Chief for the Transactions on Pattern Analysis and Machine Intelligence (PAMI) and as an Editorial Board member for the *International Journal of Computer Vision* (IJCV). She previously served as a Program Chair of the IEEE Conference on Computer Vision and Pattern Recognition (CVPR) 2015 and a Program Chair of Neural Information Processing Systems (NeurIPS) 2018. https://www.cs.utexas.edu/users/grauman/
- Trevor Darrell – Prof. Darrell is on the faculty of the CS and EE Divisions of the EECS Department at UC Berkeley. He founded and

co-leads Berkeley's Berkeley Artificial Intelligence Research (BAIR) lab, the Berkeley DeepDrive (BDD) Industrial Consortia, and the recently launched BAIR Commons program in partnership with Facebook, Google, Microsoft, Amazon, and other partners. He also is Faculty Director of the PATH research center at UC Berkeley, and previously led the Vision group at the UC-affiliated International Computer Science Institute in Berkeley. Before that, Prof. Darrell was on the faculty of the MIT EECS department from 1999 to 2008, where he directed the Vision Interface Group. He was a member of the research staff at Interval Research Corporation from 1996–1999, and received the S.M. and Ph.D. degrees from MIT in 1992 and 1996, respectively. He obtained a B.S.E. degree from the University of Pennsylvania in 1988. Darrell's group develops algorithms for large-scale perceptual learning, including object and activity recognition and detection, for a variety of applications including autonomous vehicles, media search, and multimodal interaction with robots and mobile devices. His areas of interest include computer vision, machine learning, natural language processing, and perception-based human–computer interfaces. Prof. Darrell also serves as consulting Chief Scientist for the start-up Grabango, which is developing checkout-free shopping experiences, and for Nexar, which is pioneering city-scale visual driving analytics. Darrell also recently co-founded Datu with former students and postdocs. Darrell is on the scientific advisory board of several other ventures, including WaveOne, SafelyYou, and KiwiBot. Previously, Darrell advised Pinterest, Tyzx (acquired by Intel), IQ Engines (acquired by Yahoo), Koozoo, BotSquare/Flutter (acquired by Google), MetaMind (acquired by Salesforce), and DeepScale. Darrell has also served as an expert witness for patent litigation relating to computer vision. https://people.eecs.berkeley.edu/~trevor/

- Susan Athey – the Economics of Technology Professor at Stanford Graduate School of Business. She received her bachelor's degree from Duke University and her Ph.D. from Stanford, and she holds an honorary doctorate from Duke University. She previously taught at the economics departments at MIT, Stanford, and Harvard. Her current research focuses on the economics of digitization, marketplace design, and the intersection of econometrics and machine learning. She has worked on several application areas, including timber auctions, Internet search, online advertising, the news media, and the application of digital technology to social impact applications. As one of the first "tech economists", she served as consulting chief economist for Microsoft Corporation for six years, and now serves on the boards of Expedia, Lending Club, Rover, Turo, and Ripple, as well as non-profit Innovations for Poverty Action. She also serves as a long-term advisor to the British Columbia Ministry of Forests, helping architects to implement their auction-based pricing system. She is the founding director of the Golub Capital Social Impact Lab at Stanford GSB and associate director of the Stanford Institute for Human-Centered Artificial Intelligence. https://www.gsb.stanford.edu/faculty-research/faculty/susan-athey

Although the knowledge produced by these scientists seems to be different from each other, they contribute to the emergence of high-value-added products and services on a universal scale.

2.2 ALTERNATIVE VIEWS OF DIGITAL TRANSFORMATION IN THE UNIVERSITIES

The massive increase in production has accelerated as a result of the transfer of knowledge through universities. Since the existing knowledge is constantly transformed in universities, it is reflected in practice in a way that increases production even further. Universities accelerate the growth in production processes since they internalize technology (Clarke et al. 2013). Especially, the development of digital technologies and the use of open-source software programs accelerate the application of theoretical knowledge in universities. At this point, the implicit knowledge through the strong human capital in the universities contributes to the application processes. In this way, the velocity of "implicit knowledge" becomes a determining variable in the digital transformation of the universities.

According to the results of an analysis conducted by McKinsey Global Institute (2019), it is stated that five main themes that determine a successful digital transformation process which is shown in Figure 2.4.

The digital transformation in the HE institutions creates a surplus for stakeholders via "economies of scale" regardless of the size of the university (World Bank 2016). From this point of view, the digital transformation accelerates and enlarges itself continuously with the added value it has created since its inception. However, the factors that determine the management of the digital transformation process will also determine its effectiveness.

It is important to have the ability to measure for efficiency should also take into account the determination of rules, criteria, and norms for evaluating the digital transformation. This topic is obvious in terms of the literature, and there are various studies (Stigler 1961; Aghion et al. 2001; Bresnahan et al. 2002; Acemoglu et al. 2005; Commander et al. 2011; Andjelkovic 2015) based on determining the proper sizes for scale economies.

In today's world, digital transformation seems to be an effective tool due to the opportunities provided by the economies of scale mentioned above. The main reason for this is that the positive knowledge and value transfer of people. However, the factors that will play a role in the emergence of this

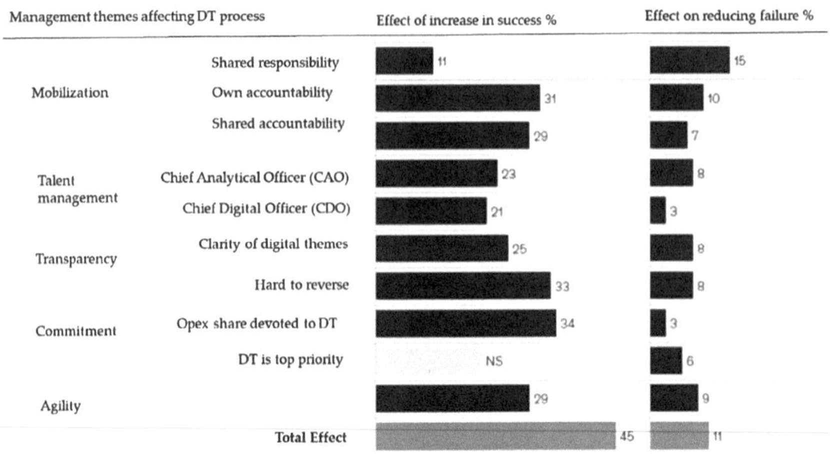

Figure 2.4 Major themes for successful digital transformation.

Source: McKinsey Global Institute (2019).

result are the basic variables of the standard economic and business process. In this framework, the management of the existing resources for digital transformation and the management process for this is much more important. Findings are showing that artificial intelligence (AI) applications do not have a significant impact on the basic resource in digital transformation. Here, McKinsey Global Institute (2019) specifies the existence of five basic themes, namely "mobilization", "talent management", "transparency", "commitment", and "agility" as a managerial process for the variables that will determine the efficiency in resource use in digital transformation (Figure 2.4).

These basic themes are divided into breakdowns within themselves. Considering the implementation of these five basic themes together with their sub-breakdowns as shown in Figure 2.4; it has been revealed that there is a 45% efficiency increase in digital transformation projects, and an 11% reduction in the effect of the failure of digital transformation projects is measured and hence, 66% efficiency level achieved in total. In order to ensure operational continuity in the digital transformation process, it is necessary to control the liquidity of the transactions. For this, a sustainable liquidity management approach should be applied for a sustainable digital transformation process. In this process, the internal audit mechanism is critical to monitor the effective management of liquidity in the HE institutions and to evaluate the attainability of strategic goals.

Prager Sealy, KPMG, and Attain (2010) have jointly stated the fact that the analysis of strategic risks which means the difference between the strategically balanced and financially balanced operating and capital budgets are

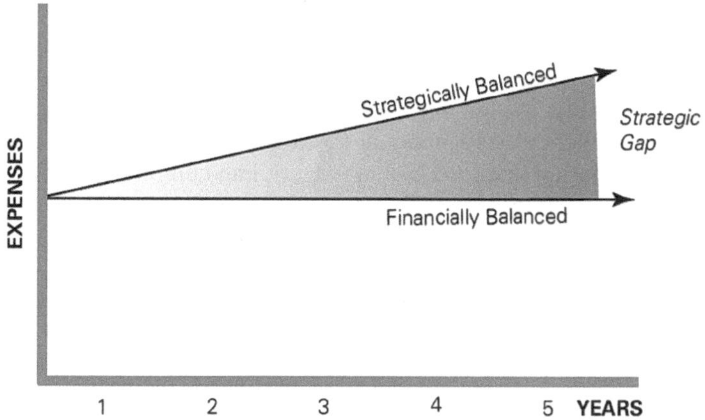

Figure 2.5 Determining strategic gap between operational and capital budgets to manage liquidity at the university.

Source: Prager Sealy, KPMG, and Attain (2010).

executed through strategic audits (Figure 2.5). It is also important to report relevant recommendations attached to the key findings to the top management can be expressed as a best practice, especially within a modern understanding of auditors being the "trusted advisors".

Sometimes, there may be a strategic gap in either capital or operating budget of the HE institution (McCaferry 2019). In any case, this situation should be a concern for the management. However, if the existence of such a strategic gap is not communicated with related parties, it leads to deviations from the strategic objectives. In this respect, the HE institution's financial health and risk management must identify strategic gaps and report them regularly. Internal auditors can act as a bridge in communicating strategic issues to management. Hence, internal auditors should consider these potential risks since this may increase the HE institution's financial risk level and reduce its risk capacity and/or tolerance level. In addition, if it is a significant parameter, then it would be used in assessing other institutional risks.

Here, the Higher Education e-business model is mainly based on process analysis. It should be noted that there may be some cases in practice where universities are applying either wholly "function-oriented" or wholly "process-oriented" management style. Nevertheless, evidence suggests that the more successful universities in the HE sectors are becoming increasingly process orientated, and managed, especially considering the customer-facing parts of the organization (Daniel 2014).

The proposition of this book is that all should benefit from understanding the key processes/business drivers/risks within a Higher Education business. In this respect, a generic e-business model is proposed and intended to be dynamic. In other words, it may be necessary to update the content of this business model periodically as the education industry evolves and Higher Education professionals add further to their knowledge of the industry. In principle, the model is intended to be of use by HE professionals who are working in a university at managerial positions, including finance managers, IT managers, internal auditors, risk managers, and compliance professionals involved in Higher Education work.

2.3 TECHNOLOGY IN THE DIGITALLY TRANSFORMING UNIVERSITY

High-quality HE institutions are decisive to individuals as well as to the labor force. In order to achieve this, it needs to become more efficient and effective. However, a remedy cannot be activated by dropping the standards for success criteria. The universal aim of the HE sectors should be based on supporting the students to gain the competencies needed to succeed in the 21st-century business world. Moreover, the education parameters should be in place to teach them how to be active citizens equipped with analytical thinking, writing, and quantitative judgment skills (Pusser et al. 2006).

The concept of the digital business model should be analyzed with an extensive approach to involve the key elements of the HE institutions' structure that define different business processes and core operations through which value is created (Hedman and Kalling 2003). Determining the principal concepts of "centralization", "formalization", "standardization", and "specialization" is essential for the HE institutions in a structural context.

Decisions are taken by the leaders in the centralized structure as opposed to those at lower levels in the hierarchy (Ireland and Webb 2007). Whereas in a decentralized organization, critical decisions are taken by the staff. In the context of HE institution, decisions are made by Deans or Directors in a centralized university, while in a decentralized case, decision-makers are the individual faculties. *Standardization* is defined as the degree to which staffs' behavior and decisions are determined by rules, regulations, and procedures that apply to all the staff (Child 1972, 1973; Ireland and Webb 2007).

In the context of higher education, standardization can be interpreted as the policies, procedures, and practices about the design of courses. For instance, preparing procedures to define when to develop a new course and the overall design. In addition, there may be a higher level of standardization which includes charters about the number of units, the kinds of assessments used, syllabi templates, etc.

In the context of higher education, standardization can be interpreted as the policies, procedures, and practices about the design of courses.

For instance, preparing procedures to define when to develop a new course and the overall design. In addition, there may be a higher level of standardization which includes charters about the number of units, the kinds of assessments used, syllabi templates, etc.

Formalization is defined as "the extent to which procedures, rules, instructions, and communication are written down" (Child 1973; Ireland and Webb 2007). Examples for formalization in the HE context are the job descriptions and different types of contracts, instructions, user handbooks, and the extent to which policies, procedures, and rules are codified and documented.

Specialization is defined as "the extent to which work is broken up into discrete jobs" (Child 1973). In the higher education sector, the work of designing and teaching courses should be done by "a professor" who is specialized in different fields.

Grajeck (2016) states that the HE sectors intend to invest in wireless access, adjust, and upgrade their teaching technologies, and determine what to do about a dated version of their enterprise applications. On the other hand, IT funding in HE sectors is still at its early stages and it is called like "zero-sum game" in the literature which will for sure leads to a strategy for "digital insolvency".

McLachlan et al. (2017) associate the digitalization process with a "glue for the university" and define it as follows: "Digital transformation is the changes associated with the complete application of digital technology in all aspects of a modern university". In this respect, Figure 2.6 indicates the difference between the digital capabilities of universities in the past and future states, respectively. It should be noted that the traditional IT department structure falls short in the new university business model.

It is important to change the IT department structures primarily during the transition period. Because this department, which constitutes the mindset of the digital transformation process of universities and has strategic

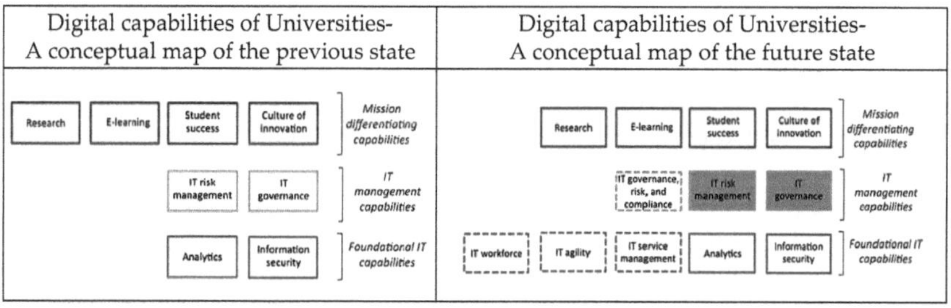

Figure 2.6 Digital capabilities of universities – a conceptual map of past versus future states.

Source: Grajeck (2016).

importance, should be composed of competent personnel and equipped with modern tools and techniques. Especially, the foundational IT capabilities play a key role in the digital transformation process to achieve a successful implementation (Grajeck 2016).

Rubin (2013) states that there are two major forces that have impacted higher education: the use of technology to teach via online learning, and the digital business models, policies, and structures of the HE institutions. While technology has the potential to increase access and efficiency, the outcomes mostly depend on its implementation style.

2.4 ON THE NATURE OF THE DIGITAL AUDIT: A BASIS FOR CRITICAL DISCUSSION ON AUDIT CULTURE

Audit, accountability, and quality assurance have been a constant consideration of the HE institutions from the early days of their establishment (Amaral 2014). In time, this becomes a public issue to maintain the trust in the HE institutions. Considering the risky business environments, the risk analysis which is set out in the process analysis is designed to be a crucial part of the model from an audit perspective. The model is designed to help HE professionals build up an understanding of the effectiveness of a university's business design and management and of the critical performance-related issues it faces. In this way, HE professionals are expected to:

1. Better evaluate the *audit universe* of HE institutions
2. Better evaluate and understand the issues and risks arising to determine the *risk universe* of the HE institutions
3. Be better prepared to make accurate recommendations by being confident in the detailed operations of HE institutions by checking the relevant *key controls* within the related workflows of each unit

It is important to understand the university's digital strategy well enough to define the audit universe, risk universe, and to execute value-added internal audit fieldwork. In this context, as a first step, it is necessary to understand the inherent risks facing the university in the virtual networks and platform structures.

The second step is to understand the design and operation of the university's internal control structure in virtual and its information systems infrastructure (both financial and operational) within the context of its' process and/or functional organizational framework.

As a third step, it is important to understand what has happened to the university's current business targets during the year as a background for benchmarking and to make an accurate gap analysis based on its results.

This is essential for generating a value-added audit recommendation preparation stage based on the future university concept.

The important point here is that the audit teams adopt an effective, efficient, and economical understanding of working and apply the audit fieldwork based on CA/CM techniques as much as possible with the help of technology (Vasarhelyi and Kuenkaikaew 2010). It is recommended that audit teams should consider the cause-effect relationship between *key risk indicators* (KRIs), *key performance indicators* (KPIs), and *key control indicators* (KCIs), correctly. In this context, they should establish an audit system for generating alerts as an "early warning mechanism" based on the strategic objectives of the HE institution. Another issue that gains importance with the effect of digitalization is *ethics* (Green Book 2017). It is essential to implement ethical principles when legal regulations are slow and ineffective in some cases (Bozkus Kahyaoglu 2019). A far-reaching endeavor for higher education institutions would be to establish business processes on the axis of KPIs, KRIs, KCIs, and Ethics. So, they can achieve their strategic goals in a globally competitive environment which is shown in Figure 2.7.

The key indicators of these axes are expressed in Figure 2.7, regarding the strategic management processes, and tabulated core business processes and

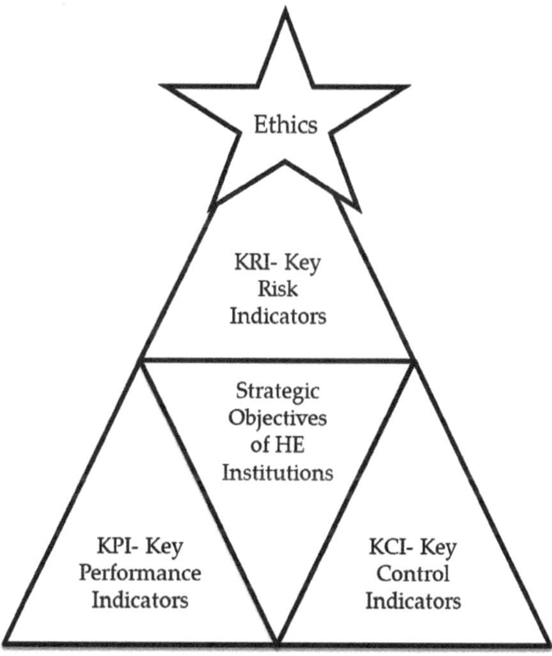

Figure 2.7 The axis of KPIs, KRIs, KCIs, and ethics in the HE institutions.

Source: COSO (2019).

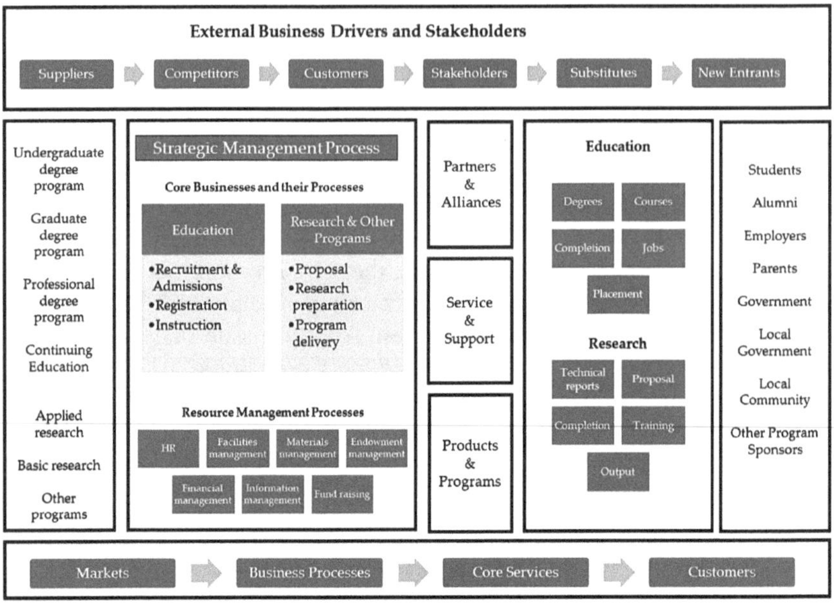

Figure 2.8 Entity level higher education business model with a holistic approach.

Source: Prager, Sealy, KPMG, and Attain (2010) and Kettunen (2015).

explained in the following sections of the summary. Please note that all the explanations are given in such a perspective to support the governance, audit, risk management, compliance, and internal control professionals positioned in the HE sectors (Figure 2.8).

2.5 EXPLORING THE DIGITALLY TRANSFORMING UNIVERSITY WITH A HOLISTIC APPROACH

There are new business models for higher education institutions that utilize online learning and new structures for teaching and learning that are recommended to cope with "disruptive innovation" in this arena (Christensen et al. 2011; Christensen and Eyring 2011; Johnson et al. 2008).

2.5.1 Entity-level business model in the HE sectors

The entity-level business model is used to define the connected operations carried out within a business entity, the external business drivers and stakeholders that support the entity, and the business relationships with people out of the entity. It should be noted that the best way to start the audit process is to define the business model at the entity level. In this way, the auditors will have a view of the big picture to determine the audit universe. Afterward, a risk-based audit plan is prepared by considering the top management's priorities and strategic objectives to support them to minimize the risks by providing reasonable assurance through audit fieldworks including physical and digital.

The major and commonly accepted components for HE sectors are indicated in the "entity-level business model" in Figure 2.6, and these components are explained with a holistic approach as follows:

External business drivers and stakeholders: Outside factors, enforcements, etc., can hinder an entity from achieving its objectives. The classification of these external forces is shown in Table 2.1, namely "business environment", "competitive environment", and "operational environment", which is defined as a "control environment" from the internal audit perspective.

Markets: Markets are generally influential for understanding an institution's position in the industry in which it operates. It is especially important to analyze the relationship between goals and strategies concerning existing segments in the market.

Business processes: The business process defined as a structured series of activities within an institution is designed to produce a specified output. A business process works as a tool for emphasizing "how work is performed rather than what is done". So, the flow of the business process is closely linked to each other. The input of one process is the output of another. Hence, internal auditors should closely monitor the structure of business processes to investigate whether the distinctive business processes are managed with specific controls.

Considering the HE sectors' structures, business processes are divided into three major groups, namely "strategic management processes", "core

Table 2.1 Higher education sector control environment

Higher education – Control environment		
Business environment	*Competitive environment*	*Operational environment*
• Political • Legal • Macroeconomics • Technological • Demographic • Social	• Competitors • New entrants • Substitute products • Buyers • Suppliers	• Markets • Customers • Competitiveness • Trade regulations • Economics

Source: Kettunen (2010).

business processes", and "resource management processes" as shown in Figure 2.6.

2.5.2 Strategic management process in the HE sectors

Strategic management processes are essential for managing the business risks, both physical and digital in HE sectors by establishing efficient business processes and monitoring the progress toward meeting the strategic business objectives (Kettunen 2008). Internal auditors should start their fieldwork by analyzing the nature of strategic management processes in the HE institutions. In this way, they could gain a fair idea about the overall direction of the university considering the control environment, and potential opportunities and threats. The description of the strategic management process, objectives of the process, evaluation factors of the process, key risks and success factors related to this process, and audit objectives linked to KPIs, KRIs, and KCIs of this process are summarized in Table 2.2.

2.5.3 Core businesses and their processes in the HE sectors

Core business processes are divided into two major categories as shown in Figure 2.6. The first one is education and the second one is research, and other programs, respectively. The education process has three sub-processes namely, "recruitment and admissions", "registration", and "instruction" which are presented in Tables 2.3, 2.4, and 2.5, respectively.

2.5.4 Research and other sponsored programs in the HE sectors

Research and other sponsored programs process in the HE sectors are divided into two major categories such that "proposals for research preparation", and "program delivery" which is shown in Figure 2.6 (Tables 2.6 and 2.7).

2.5.5 Resource management process in the HE sectors

Resource Management processes in the HE sectors are divided into seven major categories: "Fund raising", "HR management", "Facilities management", "Materials management", "Endowment management", "Financial management", and "Information management", respectively, which is shown in Figure 2.6 (Tables 2.8 to 2.14).

Table 2.2 Strategic management process

Strategic management process description	Strategic Management process is the key process for defining the mission of the university. In this respect, the society, the HE institution is willing to serve in digital, should be formalized into a mission statement. Then this statement should be converted into a strategy that identifies digital market niches and online education programs and services to be offered by the HE institution. It should be noted that the mission could be defined either in a broad view or a narrow view. In other words, the mission may be singular, i.e., teaching students only from a specific geographic area or multiple, i.e., educating students, carrying out research, etc. Similarly, the society served by the HE institution may be a local, regional, national, or international arena. In this case, there may be some differences depending on the priorities for each mission. The digital strategy of the HE institution will then be codified into a long-term plan. The long-term digitalization plans should be harmonized with short-term operating plans such as operating budgets, capital budgets, IT infrastructures, and digital curriculums with program reviews.
Process objectives	1. Promoting the brand image and reputation of the HE institution through social media, events, and meetings 2. Maintaining accreditation status of the HE institution 3. Implementing new educational programs based upon market research or strategy/mission 4. Maintaining a balanced and sustainable budget
Evaluation factors linked to process objectives	• Designing programs such as education, research, and other sponsored programs that can appeal to the digital market niche or create benefits against the competitors (**1, 3**) • Establishment of favorable actions toward HE institution by stakeholders such as unions, society, competitors, students, donors, faculty, financial resources, social media, regulators, etc. (**1**) • Ensuring the optimal level of financial, human, facilities, and IT resources to assist the digital transformation missions of HE institution (**4**) • Providing the full conformity with regulatory requirements to maintain accreditations (**2**)
KPIs linked to process	• Number of publications and research reports • Positive factors relevant for accreditation body evaluation process • Annual benchmarking reports with peer group • Satisfaction surveys regarding the external customers related to education programs, students, program recipients, and donors • Satisfaction surveys regarding the internal customers related to resource management processes and employees • Actual financial outcomes such as liquidity ratio, efficiency ratio, and the ratio of net revenues to total revenues

(*Continued*)

Table 2.2 (Continued) Strategic management process

KRIs linked to process objectives	a) Any kind of negative publicity leading to decrease applications to the HE institution (**1**) b) Competition's cost of education is lower (**4**) c) Alterations in society or resource provider needs and expectations (**3**) d) Lack of variety in revenue-generating sources (**4**) e) Lack of communication and teamwork between core business and resource management processes (**4**) f) Lack of conformity with regulations (**2**)

KCIs linked to KRIs	• Approved version by a board of trustees in use of (**a, b, c, d, e, f**)
	  • Mission Statement of HE institution
	  • Strategic Plan of HE institution
	  • Institution-Wide Budget of HE institution
	  • Capital Budgets
	  • New Educational Programs
	  • Fees for Providing Educational Programs
	  • Information Technology Plan
	  • Campus Master Plan (Physical & Digital)
	  • Significant Corporate Governance Policies (conflict of interest, compensation programs), and
	  • Transactions and ethical culture (executive compensation, large gifts)
	• Systematic and periodic Benchmarking of results against the strategic plan (**a, b, e**)
	• Benchmarking against peer group (**b, c, d**)

Audit objectives	• Analyzing the overall business risk, leadership style, management approach, and control environment • Identifying material financial accounting issues,[6] auditings, reporting significant issues, and planning target audit procedures accordingly
Audit test procedures	• The major audit approach for this process will be basically the inspection of the control environment both physical and digital and tests of strategic controls, respectively • Internal auditors should determine which of the previously defined key management controls significantly have an impact on the control environment and test them

Source: Adapted by the Authors from IIA (2017); Prager Sealy, KPMG, and Attain (2010); Kettunen (2015); Kandeh and Alsahli (2020).

Table 2.3 Recruitment and admissions process

Recruitment and admissions process description	• In this process, Education Programs and the HE institution as a whole are promoted to target groups and potential students. Afterward, applications are collected and reviewed. Based on the HE institution's criteria, selections are made for acceptance, and acceptance letters are sent to the accepted students. Finally, acceptance forms and deposits from students are processed
Process objectives	1. Lowering the cycle time to process students by X hours 2. Creating and maintaining a pool of applicants to the HE institution at X times students admitted 3. Accepting students in the top X% of their class and with X score in SAT or ACT 4. Increasing the admissions from nontraditional students by Y% from the prior year 5. Increasing the total registration by X% 6. Increasing the profit from accepted students by X%
Evaluation factors linked to process objectives	• Identifying the target population (**2**) • Performing the marketing plan (**2, 3, 4, 5, 6**) • Informing speedily and adequately with entrants (**1**) • Attracting targeted population (**2, 3, 4, 5**) • Providing education programs at a reasonable fee (**2, 3, 4, 5, 6**) • Providing a competitive economic aid package (**2, 3, 4, 5, 6**) • Providing a required academic program and faculty (**2, 3, 4, 5, 6**)
KPIs linked to process	• Number of days between applications received and notification of acceptance • Direct cost per student related to the admission process • Marketing costs per incoming freshman • Marketing costs per applicant • Number of applicants, admitted students, and matriculants with specific criteria • Admitted students per average number FTE-Full-time equivalent professional staff • Ratio of admittances to applicants • Tuition and fees compared to the peer group • Number of transfers to/from the HE institution • Number of inquiries this year vs. prior years • Maintain SAT or ACT scores within a range that represents the students targeted by the strategic plan of the HE institution

(*Continued*)

Table 2.3 (Continued) Recruitment and admissions process

KRIs linked to process objectives	a) Variations in number and type of potential students (**2, 3, 4, 5, 6**) b) Inoperative marketing approach (**2, 4, 5, 6**) c) Quality rankings prepared by third parties decreased from the prior year (**2, 4, 5, 6**) d) Competitors with lower fee/better education programs (**2, 4, 5, 6**) e) Shift in the market for skills on which education programs are based (**2, 4, 5, 6)** f) Compromising quality by growing in number of admitted students (**3**) g) Tuition costs lower than that of competitors (**2, 4, 5, 6**) h) Inability to offer competitive economic aid Controls (Operational, Financial, Compliance) (**4, 5, 6**) i) Inefficient work style of resourcing in admission department which slows down cycle time (**I**)
KCIs linked to KRIs	• Monitoring the status of each application received and responded (**a, b, e, f**) • Comparing the number of applicants, admissions, retainage, and matriculants based on the plan (**a, b, e, f**) • Comparing the tuition and all fees to the peer group (**d, g, h**) • Follow-up on accepted applicants that chose another HE institution (**b, c, e**) • Comparing the admitted students against peer group (**e**) • Comparing the accepted students against admittance standards (**f**) • Responding time regarding the applicants receive of accept/reject information (**i**)
Audit objectives	It should be noted that the key controls and KPI s over this sub-process are critical since they will underline the changes that will take place in the future, such as decreases in the number of applications, admitted students. This may lead to a lower quality of enrollments that may affect the collection of accounts receivables[7]
Audit test procedures	• Internal auditors should determine which of the previously defined key controls significantly reduce the risk of misstatement and test them accordingly • There is a need for "Substantive Procedures" such as performing analytics by using applicable KPI s; applying "Trend analysis" of admission deposit fee revenue and comparing the fee revenue to applications received

Source: Adapted by the Authors from APQC (2014); Arcuria (2015); IIA (2017); Kandeh and Alsahli (2020).

Table 2.4 Registration process

Registration process description	In this process, students may apply for financing (financial aid). The major steps of this process are registration, granting of financial aid, payment plans, billing, and cash receipts. Sometimes, housing services are also included
Process objectives	1. Increase in online registration or preregistration by X% 2. Decrease in average waiting time for online registration by X hours 3. Reduction in the number of credit hour drop/adds by X credit hours 4. Increase in utilization of housing by X% 5. Increase in the minimum students registered for courses 6. Providing students with X% of first choice courses 7. Providing students access to all courses required for graduation within a certain period 8. Decrease in the number of courses with less than minimum size to X 9. Increase in the online course utilization to an average of X% 10. Increase in the percentage of tuition collected prior to the start of classes by X%
Evaluation factors linked to process objectives	• Assisting the students conveniently and cost-effectively (**2**) • Effective use of technology to register students (**1**) • Accurate internal communication among all departments for interacting with students (**3, 10**) • Providing relevant digital curriculum content and time that satisfy students' needs and expectations (**3, 5, 6, 7, 8, 9**) • Place all requesting students in university housing (**4**) • Establishing a billing system to serve on a timely basis (**10**)
KPIs linked to process	• Average process time for students to register • Number of credit hours drop/add • Number of students placed in courses they want • Number of beds occupied as compared to available • Number of students on meal plan compared to full-time students • Average tuition and fees per student • Average financial aid to tuition and fees per student • Unfunded aid to tuition and fees

(*Continued*)

Table 2.4 (Continued) Registration process

KRIs linked to process objectives	a) Technology failure which leads to slow processing (**1, 2**) b) Available courses or faculty not satisfying the needs and expectations of students (**3, 6, 7, 8, 9, 10**) c) Inappropriate pricing strategy (**4**) d) Unbilled or uncollected tuition revenue (**10**) e) Students dissatisfied with institutions' housing facilities (**5**) f) Higher costs due to disproportionate level of course/student/faculty levels (**6, 9, 10**)
KCIs linked to KRIs	• Preparing student surveys (**b, f**) • Monitoring the complaints and inspections of the process performed by senior management (**b, c, e, f**) • Applying general IT controls (**a, d**) • Implementing effective credit policies and payment plans (**d, e**) • Monitoring the unfunded financial aid (**b, d**) • Analyzing the budgeted tuition revenue to actual enrollment statistics (**d**) • Monitoring competitors' pricing strategies (**c, f**) • Implementing appropriate course offerings and size to registration statistics (**b, c, f**) • Effective communication and coordination among departments (**b, e**)
Audit objectives	Considering routine transactions in the registration process, it is recommended for the auditors to use a systems-based approach. It should be noted that the non-routine transactions are rare and most of the time consist of government grants per student. Hence, internal auditors can apply accounting estimates regarding the allowance for doubtful student accounts. Auditors should consider analyzing the following accounts and transactions[8] related to the registration process: "Cash", "Student Accounts Receivable", "Allowance for Doubtful Accounts", "Student Loans Receivable", "Tuition and Fees Revenue and Expense", "Auxiliary Enterprise Revenue", and "Scholarships and Financial Aid Expenditures"
Audit test procedures	In this respect, the Audit Objectives related to the financial statement accounts include the following items: • Convenient allowance for uncollectable amounts of student accounts receivable and student loans receivable. In addition, government-funded loans should be analyzed separately, since the funding of loan losses may be different (EXISTENCE, ACCURACY, VALUATION, AND COMPLETENESS) • All tuition and fee revenues are properly recorded (COMPLETENESS AND ACCURACY)

Source: Adapted by the Authors from Prager Sealy, KPMG, and Attain (2010); Arnold and Pistilli (2012); APQC (2014); IIA (2017); Kandeh and Alsahli (2020).

Table 2.5 Instruction process

Instruction process description	The most important process of the Education Core Business is the instruction of the students, educating them, providing necessary vision and directions to them, finally graduating them, and placing them with employers or higher levels of education. In addition, students are supposed to purchase related books and other course materials during this process
Process objectives	1. Growing or maintaining the retention and graduation levels at a specified percentage 2. Assistance to place X% of students among employers prior to graduation 3. Increasing the level of satisfaction regarding the course contents and faculty 4. Increasing the grade points on the average per student body by X 5. Improving the rating for faculty undergoing instructional peer review 6. Maintaining the conformity with accrediting agency standards
Evaluation factors linked to process objectives	• Establishment of academic programs that satisfy the institutional and accrediting agency standards (**6**) • Increasing the level of student satisfaction (**1, 3**) • Satisfying the educational needs and expectations of employers and society in such a way that provides the students with employable skills or advances their education (**1, 2**) • Keeping the class size at an acceptable and competitive level (**1**) • Maintaining the quality of faculty (**3, 4, 5**) • Using advanced technology in instruction (**2**) • Providing suitable, well-maintained facilities and digital network platforms (**1**)
KPIs linked to process	• Faculty cost level per student • Contribution level by program • Student FTE by course • Faculty-to-student ratio • Average GPA of graduating class • Ranking of programs at national/local level • Faculty scholarly works published or awards to faculty/full-time faculty • Student surveys related to faculty and education programs • Acceptance rate into graduate schools • Graduates employed and average salaries per graduate • Employer satisfaction with graduates hired

(*Continued*)

Table 2.5 (Continued) Instruction process

KRIs linked to process objectives	a) The needs and future expectations of employers for skills of students dramatically change due to digital transformation (**2**) b) Students are not eligible enough to attend graduate schools (**1**) c) Education services provided by the HE institution do not meet the needs and expectations of students (low quality, high fee) (**3**) d) Dissatisfied students leave university (**1, 3**) e) Dramatic change in the accreditation criteria due to digitalization (**6**) f) Lack of sufficient faculty expertise due to digital transformation (**5**) g) Ineffective career placement office (**2**) h) Inappropriate class size preventing faculty to be effective (**4, 5**)
KCIs linked to KRIs	• Conducting and reporting student and alumni surveys (**c, d**) • Benchmarking of graduation rates versus employment rates (**a**) • Monitoring the appropriate size of the class (**h**) • Monitoring the activities of career placement office (**g**) • Monitoring the accreditation criteria on an annual basis (**e**) • Conducting and reporting surveys with employees (**a**) • Conducting student–advisor sessions to improve the progress of students (**b**) • Monitoring instructional peer review and providing necessary training (**f**)
Audit objectives	It should be noted that the actual instruction of students mostly includes the transactions generated at the resource management processes such as instruction payments to faculty, maintenance of the facilities for instruction. In addition to this, obtaining supplies and materials for instruction should be considered. Testing these transactions can be actualized in the appropriate level of resource management sub-processes
Audit test procedures	Internal auditors should determine which of the previously defined management controls reduce the risk of material misstatement and test them In addition, they can use "Substantive Procedures" such as performing data analytics by using applicable KPI s, using trend analysis of fee revenue, comparing the fee revenue to the number of graduates

Source: Adapted by the Authors from Prager Sealy, KPMG, and Attain (2010); Arroway et al. (2016); IIA (2017) and Kandeh and Alsahli (2020).

Table 2.6 Proposal for research preparation process

Proposal for research preparation process description	The core business of the research process covers various activities related to financing research projects by external funds. There are three major fund-providing parties, namely governmental units, private institutions, and private foundations. These parties mainly provide funds to the HE institution to support institutional research and other sponsored activities identified by the resource provider or society that are consistent with the HE institution's mission It should be noted that the sponsored research programs require stewardship and accountability in the governance of the funds on behalf of the external sponsors. In this respect, this process involves key project development activities consisting of proposal preparation and submission
Process objectives	1. Decreasing and maintaining the average response time to proposals by X number of days 2. Increasing the success rate of proposals by X% 3. Increasing the budget in research per faculty member by $Z 4. Increasing the research base by X% 5. Increasing the funding for departmental research by private entities by X% 6. Ensuring the preparation of proposals in-line with the institution's mission only
Evaluation factors linked to process objectives	• Hiring and keeping technical staff with high competency levels (**1, 2, 3, 4, 5**) • Responding speedily to the relevant research opportunities in the market (**1**) • Diversifying the number of resource providers to achieve more research projects (**4, 5**) • Diversifying the number of different grant programs (**4**) • Maintaining and improving the quality of research programs and research facilities (**4, 5**) • Being more selective to focus on research that is obtaining funding (**2, 3, 4**) • Ensuring the review of proposals prior to submission (**6**)
KPIs linked to process	• Total number of man-hours regarding the proposal preparations • Projected research base and funding source • Total number of proposals prepared versus the accepted proposals
KRIs linked to process objectives	a) Loss of proposals against competitive parties (**2, 3, 4, 5**) b) Reduction in the levels of funding from sponsor sources (**3, 4, 5**) c) Losing key research staff (**1**) d) Generation and acceptance of proposals that are under-scoped (**3, 4**) e) The potential risk of accepting proposals inconsistent with the HE institution's mission or which may lead to negative public reaction (**6**)

(*Continued*)

Table 2.6 (*Continued*) Proposal for research preparation process

KCIs linked to KRIs	• Assigning new private sources of funding for research programs (**b**) • Monitoring the grant proposals submitted to ensure the quality standards (**a, d, e**) • Reviewing the grant proposals by knowledgeable personnel to ensure the quality standards (**a, d, e**) • Monitoring the budgets over overhead items (**b**) • Monitoring the key staff utilization (**c**)
Audit objectives	It should be noted that the financial transactions in this process are mainly related to the submission of proposal reports to the relevant resource providers to obtain recovery of overhead and fringe benefits for proposals submitted. On the other hand, the actual transactions for research activities will be usually occurring in the delivery process. In this respect, internal auditors should apply the tests in relation to the key controls to determine whether faculty management is monitoring and administering the proposal function. This information will give an idea for the auditors in evaluating financial results on a long-term basis
Audit test procedures	Internal auditors should apply substantive procedures in this process. It should be noted that the financial statement items affected by this sub-process are indirect cost revenues and direct cost revenues considering the fringe benefits. In addition, this sub-process is related to an audit program consisting of compliance with regulations

Source: Adapted by the Authors from Prager Sealy, KPMG, and Attain (2010); Aziz et al. (2012); APQC (2014); IIA (2017) and Kandeh and Alsahli (2020).

Table 2.7 Program-delivery process

Program-delivery process description	This sub-process mainly involves the delivery of sponsored research activities to third-party resource providers. In addition, please note that there may be billing and collecting costs incurred
Process objectives	1. Increasing the number of patents compared to the prior year by X 2. Minimizing the cost overruns funded by the HE institution by $Y 3. Decreasing the total days between expense incurrence and drawdowns from the government by X 4. Increasing the submitted technical reports on time by Y% 5. Increasing the publication activity based on the research
Evaluation factors linked to process objectives	• Submission of all the deliverables within defined deadlines (**3, 4**) • Closing the projects within budget constraints and related regulations (**2**) • Providing adequate research facilities (**1**) • Successfully achieving the grant objectives (**1**) • Hiring and retaining key research staff with technical competency (**5**)
KPIs linked to process	• The total number of technical reports prepared on a timely basis • The total number of research in dollars amount per faculty • The total amount of cost overruns related to the grants • The total amount and the number of cost transactions not approved • The total number of the patents • The total number of articles published in relation to the research • The research database by sponsor and department
KRIs linked to process objectives	a) Funding denied due to unsuitable technical product (**1, 5**) b) The problems arising from the sub-recipient/subcontractor whose works are not according to specifications/timely (**4**) c) Rejecting the costs due to noncompliance with regulations (**2**) d) Economic crisis and volatile conditions affecting funding sources (**1, 5**) e) Lack of review of grant expenses by management which is needed for preparing the reimbursement forms (**3**) f) Due to time constraints, key research staff is not able to prepare publications (**5**) g) Lack of review of the project budget (**2**)

(*Continued*)

Table 2.7 (Continued) Program-delivery process

KCIs linked to KRIs	• Peer-review process applied for the reports submitted (**e**) • Monitoring the status of technical reports (**a**, **b**) • Monitoring the performance of sub-recipient/subcontractor (**b**) • Monitoring the compliance with regulations by appropriate personnel (**c**) • Comparing the costs incurred versus project status or budget (**e**, **g**) • Identifying the multiple sources of funding for research programs (**d**) • Monitoring the utilization of the key research staff (**f**)
Audit objectives	It should be noted that the key financial transactions generated in this process are mostly related to the recognition of direct and indirect cost revenue, submission of billings to the sponsor, and collection of receivables. Hence, the accounting estimates include the valuation of accounts receivable either billed or work-in-process. The routine transaction types of this process are defined as the recognition of direct cost revenue, recognition of indirect cost revenue, billing of costs to third parties and expenditures for salaries and benefits, and other items related to the research grants. In this respect, internal auditors should consider analyzing the allowances for doubtful receivables related to the sponsored programs
Audit test procedures	It should be noted that this sub-process is usually audited indirectly because revenue is recognized when the related costs are incurred. Internal auditors should confirm the process that if the costs are legal and accurate, then the revenue recognition is also accurate. The transactions to recognize revenue are generally automated in the HE institutions, and they are mostly dependent on the costs incurred. Hence, it is recommended to apply a systems-based approach (CA/CM) for auditing.

Source: Adapted by the Authors from Prager Sealy, KPMG, and Attain (2010); APQC (2014); IIA (2017) and Kandeh and Alsahli (2020).

Table 2.8 Fundraising process

Fundraising process description	It should be noted that this process covers getting financial resources from various sponsors to be used for education, long-term investment, or construction/acquisition of facilities. In this respect, the fundraising process should be in line with the HE institution's vision, mission, and strategic plan. The key sub-processes of the fundraising are conducting market research to determine the potential sponsors, the establishment of sponsor programs related to the alumni activities, fundraising programs, and sponsor products namely types of contributions such as deferred giving, gifts of property. The potential sponsors are found and established close communication in order to organize the receipt of a pledge to make a contribution. They executed any necessary documents, sent pledge acknowledgments, billing notices, and cash receipt notices. There may be attempts to find potential future sponsors.
Process objectives	1. Increasing the sponsor aid by X% 2. Increasing the alumni giving by X% 3. Increasing the alumni participation by X% 4. Increasing the amount of revenue by X per employee 5. Reducing the time from pledge to receipt by X days 6. Reducing the amount of money spent to raise new funds
Evaluation factors linked to process objectives	• Establishment of alternative sources of revenue/support (**1**) • Promoting the existing sources of revenue/support (**1**) • Hiring and retaining competent and knowledgeable staff in fundraising techniques (**1, 2, 3, 4**) • Establishment of effective monitoring system and database management regarding the potential sponsors and alumni (**1, 2, 3**) • Collection of pledges made (**5**) • Satisfying the donor/alumni's needs and expectations (**1, 2, 3**) • Promoting and strengthening the sponsor acceptance process (**1**) • Involving the HE institution's management in the fundraising process (**1, 2, 3, 4**) • Establishment of decent reputation and recognition (**1, 2, 3, 4**) • Creating efficient, effective, and economic fundraising efforts (**6**)

(*Continued*)

Table 2.8 (Continued) Fundraising process

KPIs linked to process	• Total number of days between amount pledged and cash received to the HE institution • Average time spent for finding a sponsor • The percentage of alumni revenues generated within the total revenues • Total revenue amount per alumni • Gross margin per fundraising program • Average gift amount received per fundraising program • Average pledge revenue amount per employee • Total number of contacts in the fundraising program • Number of new sponsors in the fundraising program • Number of pledges received per contacts made • Total number of alumni participations • Total number of contacts per fundraising staff
KRIs linked to process objectives	a) There may be a change in related accounting and tax regulations (**1, 2, 3, 4**) b) There may be competition for sponsors among other HE institutions (**1**) c) Incompetent technical staff or ineffective fundraising campaigns (**6**) d) Weak economic conditions (**1, 2, 3, 4**) e) There may be difficulty in collecting the pledges (**5**) f) Unsatisfied sponsors (**1**) g) Lack of participation of HE institution Board members (**1, 2, 3, 4**) h) Adverse publicity (**1, 2, 3, 4**)
KCIs linked to KRIs	• Defining the strategic objectives regarding the fundraising approach of the HE institution, which reflect the state of the economy and prevent any negative publicity (deans, department chairs, program services) (**b, c, d, f, g, h**) • Monitoring the related changes in taxation and accounting rules and regulations (**a**) • Comparing the fundraising program goals and costs periodically to actual results (**c**) • Continuously monitoring results of fundraising efforts by the fundraiser (**c**) • Monitoring the pledges received, billing sent, and cash collected (**e**) • Monitoring the pledges not received (**e**)

(Continued)

Table 2.8 (Continued) Fundraising process

Audit objectives	Internal auditors should assess the validity of Receivables (EXISTENCE AND OWNERSHIP) Revenue recognition is done properly (COMPLETENESS AND ACCURACY) Considering the uncollectable amounts, adequate allowance should be provided accordingly (VALUATION, COMPLETENESS, EXISTENCE, AND ACCURACY) Internal auditors should review the transactions and ensure that Contributions Receivable and Contributions not recorded are properly disclosed in accordance with generally accepted accounting principles (PRESENTATION AND DISCLOSURE) Receivables are valued at their net present value in accordance with IFRS (VALUATION AND ACCURACY)
Audit test procedures	It should be noted that most of the transactions are routine in this process. Internal auditors should consider applying a systems-based auditing approach (CA/CM) for these transactions. On the other hand, there may be some non-routine transactions such as significant gifts received. In this respect, it is recommended to apply accounting estimates related to the review of the allowance for doubtful gifts receivable. If the HE institution applied accounting principles on a cash basis, then the accounting estimate would not be applicable

Source: Adapted by the Authors from Prager Sealy, KPMG, and Attain (2010); APQC (2014); IIA (2017) and Kandeh and Alsahli (2020).

Table 2.9 HR management process

HR Management process description	It should be noted that this process involves the human resource requirements of HE institutions in all dimensions. In this respect, considering the digital transformation process, analyzing the market of available resources is gaining more importance as a strategic issue. In addition, there are various sub-processes related to operational and compliance activities in HR management process such as institute's programs for compensation, fringe benefits, recruitment, training, and advancement. This process also integrates human resource needs and expectations with other organizational planning to achieve objectives
Process objectives	1. Hiring and retaining faculty with high visibility, competency, and analytical skills to head new "X" education programs 2. Controlling the increase in salary and fringe benefits of staff to X% overall 3. Maintaining the minimum credit hours and/or managing $X of research programs per faculty member 4. Limiting the undesirable turnover level by X% over the prior year 5. Maintaining the X% of tenured faculty to total faculty 6. Training and assisting the technical staff 7. Having full conformity with regulatory/tax filing requirements
Evaluation factors linked to process objectives	• Recruiting and hiring highly skilled and qualified faculty (**1**) • Organizing effective training programs for technical staff (**6**) • Establishment of a competitive cost-beneficial compensation package for the staff (**2, 4, 5**) • Actions to retain staff (**4, 5**) • Optimizing the utilization and productivity of employees/faculty (**3**) • Maintaining the faculty/staff commitment to HE institution (**4, 5**) • Recruiting and hiring qualified HR manager and staff (**7**)
KPIs linked to process	• Turnover rates and the employment duration • Average revenue per employee (e.g. credit hours generated per FTE faculty or departmental revenue divided by departmental payroll) • Trend analysis regarding the fringe benefits • Payroll costs by job status or department (education program) per total payroll costs • Student-to-faculty ratio • Comparing the compensation packages of the HE institution to the HE sectors • Satisfaction surveys applied for staff (compensation & benefit programs, quality of service, turnaround time) • The ratio of tenured faculty to total faculty

(*Continued*)

Table 2.9 (Continued) HR management process

KRIs linked to process objectives	a) There may be less productive tenured faculty members (**3**) b) There may be an undesired level of faculty turnover (**4, 5**) c) There may be some work stoppages or absurd salary requests due to unreliable labor relations (**2**) d) There may be some penalties or loss of tax-exempt status due to noncompliance with the rules and regulations (**7**) e) Compensation packages may be unsatisfactory for the staff (**1, 4, 5**) f) There may be some ineffective training programs (**6**)
KCIs linked to KRIs	• Controlling the utilization reports (**a**) • Reviewing the resource allocation system to ensure completeness, timeliness, and accuracy of all efforts and establishment of management reporting system (**a, d**) • Reviewing the bids by various health care and other fringe providers to achieve the best prices (**e**) • Staff satisfaction surveys (**b**) • Preparing financial analysis reports by comparing the salary expense to revenue, considering budget vs. actual and trend over time (**a**) • Comparing the salary costs to HE sectors norms (**c, d**) • Reviewing the performance evaluations for indications of poor performance (**a**) • Controlling all the fringe benefit plans and payroll tax return filings by the due date (**d**) • Determining the claims to health plans for maintaining the cost-effectiveness (**e**) • Approving the executive compensation plans by the board of trustees (**e**)
Audit objectives	• Internal auditors should review the process to ensure that unpaid amounts at the period end are recorded (COMPLETENESS) • All the recorded amounts and transactions should be accurate in relation to the services rendered (EXISTENCE AND ACCURACY) • All the recorded amounts and transactions should be allocated soundly (ACCURACY) • The cost of benefits regarding the staff should be appropriate, and the corporate principles should be applied consistently and adequately disclosed (VALUATION, PRESENTATION, AND DISCLOSURE) • Contingencies and HR commitments related to the tax and regulatory issues are properly disclosed (PRESENTATION AND DISCLOSURE)

(*Continued*)

Table 2.9 (Continued) HR management process

Audit test procedures	Internal auditors should consider the fact that the salary and fringe benefit costs are the most significant type of costs for the HE institutions. In this respect, there are financial risks, operational risks, and legal risks in the HR Management Process entirely since this process is significantly affected by the amount and complexity of government rules and regulations regarding the payroll and fringe benefit costs. Most of the transactions in this process are routine. Hence, HE institutions generally implement monitoring controls and preventive control activities. It should be noted that the types of controls may change depending on the nature of costs such as incurring costs and allocating costs. The audit approach should be a systems-based approach (CA/CM) for routine transactions. Internal auditors should apply accounting estimates by using the estimates of accrued salaries and wages, accrued fringe benefit costs, accrued vacation pay, and also the health care and worker's compensation costs due to the nature of the plans with insurance companies

Source: Adapted by the Authors from Prager Sealy, KPMG, and Attain (2010); APQC (2014); IIA (2017) and Kandeh and Alsahli (2020).

Table 2.10 Facilities management process

Facilities Management process description	This process is related to the needs of HE institutions regarding facilities, and determination of these facilities in such a way to decide how best to acquire them oversees their design and construction, and operates and maintains them. In addition, this process governs the space assignments and utilization and the risk management program over a physical plant. In this respect, this process should integrate the facility needs, planning, and acquisition with other HE institutional planning processes. On the other hand, the nature of this process is changing due to the digital transformation process since there will be no need for physical classrooms in the future and it is enough to form virtual classrooms at the virtual HE institutions
Process objectives	1. Controlling the average cost of construction to $X per square foot 2. Decreasing the number of deferred maintenance projects by X from the prior year 3. Decreasing the project overruns by X% 4. Increasing the number of projects completed within deadlines 5. Reducing the number of health and safety violations by X from the prior year 6. Reducing costs to operate multiple campuses by X% 7. Reducing excess capacity by X% 8. Maintaining and operating the physical and virtual platforms to satisfy the faculty needs
Evaluation factors linked to process objectives	• Performing effective and efficient programs in deferred maintenance (**2**) • Planning both physical and virtual space utilization (**7, 8**) • Controlling design and architecture activities both physical and virtual space (**1, 3, 4**) • Maintaining a safe and secure workplace both at physical and virtual space (**5**) • Maintaining and operating all the facilities effectively and efficiently (**6, 7, 8**) • Eliminating the costly services to transform digital (**6, 7, 8**)
KPIs linked to process	• Change in occupancy costs regarding the HE buildings, campus, and square footage • Costs to build up/install per square footage • Backlog of deferred maintenance costs • Cost of operating facilities internally versus outsourcing • Total expenditures related to plant maintenance • Plant maintenance costs as a percentage of total operating costs or total costs • Use of individual classrooms per hour, per day both in digital and physical • Square footage by function (research, administration, classroom)

(*Continued*)

Table 2.10 (Continued) Facilities management process

KRIs linked to process objectives	a) There may be a lack of capacity for potential students or faculty and hence, they may choose another HE institution due to facilities condition or a lack of space (**7**) b) There may be a lack of contingency plan or loss of property and other monetary exposures due to inadequate insurance coverage (**2**) c) There may be penalties for non-conformity with physical or digital environmental regulations (**5**) d) HE institution may lack facilities or do not have facilities fitting the current/emerging educational needs (**6, 7, 8**) e) There may be a low level of enrollment which is affecting the ability to use HE institution's space and capacity optimally (**7, 8**) f) Installment and setup costs may be too high (**1, 3**) g) There may be a lack of funding for maintenance operations (**2**) h) There may be a lack of oversight of facilities management activities and processes (**4**)
KCIs linked to KRIs	• Vendor research in the market (**a**) • Controlling facilities plans and respective capital budgets (**e**) • Controlling the deferred maintenance, renewal and replacement costs, and the condition of facilities and setting priorities for repairs and renovations (**h**) • Environmental and safety reviews performed by technical staff or professionals to prevent violations of physical safety and virtual security procedures (**c**) • Confirming and controlling the original design and architecture cost and subsequent change orders (**g, i**) • Controlling and comparing the related costs to operating budgets (**g**) • Controlling the utilization level compared to the plan (**a, e, f**) • Comparing costs of maintaining/operating facilities versus outsourcing (**h**) • Conducting the insurance reviews on annual basis (**b**)

(Continued)

Table 2.10 (Continued) Facilities management process

Audit objectives	Internal auditors should investigate the existence of all assets (tangible and intangible) and are owned by the HE institution (EXISTENCE AND OWNERSHIP) All expenses related to plants and assets should be recognized (COMPLETENESS AND ACCURACY) Contingencies and commitments of the HE institution should be recorded and disclosed properly and on time (PRESENTATION, DISCLOSURE, VALUATION, EXISTENCE, AND ACCURACY) The amounts of the assets should be capitalized appropriately (ACCURACY) The depreciation entries should be based on the accounting standards and conducted in an appropriate way of estimation regarding the useful lives and methods (VALUATION, PRESENTATION, AND DISCLOSURE) All unpaid amounts at the period end should be accrued properly (COMPLETENESS, EXISTENCE, OWNERSHIP, AND ACCURACY)
Audit test procedures	All types of facilities are considered as a significant component of resources of a HE institution. In this respect, it is important to establish a convenient facilities management process to attract and maintain high-quality faculty, staff, students, and education program recipients for the acquisition, operation, and maintenance of these resources. Similar to the other resource management processes, the transactions in this process are mostly routine. On the other hand, architecture or acquisition of buildings is generally a rare operation but such operations consist of routine cash disbursements in time. In practice, HE institutions generally apply adequate controls in the over acquisition and construction of new facilities and operation of existing facilities. However, they usually do not have adequate controls over the capital budgeting process. There may be some non-routine transactions in the HE institutions consisting primarily of execution of leases, sales of buildings, and acquisition of buildings. Internal auditors should apply accounting estimates consisting primarily of depreciation of fixed assets and estimates of accrued costs for risk management and environmental claims

Source: Adapted by the Authors from Prager Sealy, KPMG, and Attain (2010); APQC (2014); IIA (2017) and Kandeh and Alsahli (2020).

Table 2.11 Materials management process

Materials Management process description	In this process, HE institutions' needs for materials, supplies, and equipment are assessed regularly and determined how best to acquire them, review their utilization, and protect them
Process objectives	1. Decreasing the number of days between material order and receipt by X 2. Increasing the usage of surplus equipment by X% 3. Decreasing the average cost per item by $X 4. Making procurements based on large vendor contracts to create efficiency 5. Negotiating vendor discounts (net 10, etc.) 6. Establishing centralized procurement departments to achieve supply chain efficiency 7. Providing a secure and safe area for storage of supplies and materials
Evaluation factors linked to process objectives	• Planning and making projections for material and equipment needs of the HE institution accurately (**2**) • Preparing and conducting competitive bidding procedures (**3, 5**) • Producing well-established vendor relationships and in this way taking part in large vendor contracts (**4**) • Forming and maintaining an effective and accurate property management system in the HE institution (**2, 6**) • Minimizing materials costs/maximizing quality (**3, 4, 5**) • Minimizing the lead time between the order of materials and delivery (**1**) • Improving efficiencies of the procurement process (**1, 4, 6**) • Establishing a central purchasing function to increase efficiency (**6**) • Creating alarms, alerts, and locked, restricted areas for storage (**7**)
KPIs linked to process	• Average time span for processing the transactions between order time to delivery time • Percentage of discounted orders compared to total orders • Total costs and amounts of the surplus equipment • Percentage of goods on hand versus the actual needs • Total costs incurred by each vendor • Total costs to operate and sustain inventory compared to the outsourcing/just-in-time delivery • Total number of items returned
KRIs linked to process objectives	a) There may be a lack of suppliers providing the specific equipment (**3, 5**) b) There may be fraudulent cases/thefts related to the supplies or equipment (**7**) c) Rejection of the costs due to non-conformity with funding regulations (**3**) d) There may be extra payments for materials (**3, 5**) e) There may be unreliable forecasts for needs (**2, 4**) f) There may be a lack of control over the open purchase orders (**1**) g) Decentralized and virtual campus/departments (**6**)

(Continued)

Table 2.11 (Continued) Materials management process

KCIs linked to KRIs	• Research on available suppliers in the market (**a**) • Establishment of relevant and updated procurement policies to support the staff for achieving the competitive bidding or vendor discounts with the lowest possible cost (**d**) • Comparing the costs to purchase versus outsourcing and just-in-time delivery (**e, f**) • Controlling the operating budget for all materials and supplies expenses considering each functional expense type (**d**) • Providing a periodic inventory of equipment and supplies for the HE institution (**b**) • Hiring and maintaining staff with technical knowledge on the process (**c**) • Establishment of corporate culture for supporting the effective policies and procedures and communication links between multiple purchasing/receiving units (**g**)
Audit objectives	• Cash disbursements should be controlled for validity and reviewed to confirm that they are properly recorded (EXISTENCE AND ACCURACY) • Internal auditors should review all unpaid amounts at period end and confirm that they are all accrued properly (COMPLETENESS, EXISTENCE, ACCURACY, AND OWNERSHIP) • Internal auditors should consider all kinds of amounts which are not expensed and confirm that they are expected to provide future benefits (EXISTENCE, ACCURACY, VALUATION, OWNERSHIP, PRESENTATION, AND DISCLOSURE) • Internal auditors should monitor the process regarding the accounting principles to ensure that transactions are all appropriate and applied consistently (PRESENTATION AND DISCLOSURE)
Audit test procedures	The materials management process involves a major part of the HE institution's operational supporting process and expenses. It should be noted that most of the transactions are routine. There might be certain types of transactions in this process which are non-routine transactions, including the negotiation of large vendor contracts, accounting estimates, the depreciation of equipment, or valuation of inventory. Internal auditors should review all routine transactions using a systems-based approach (CA/CM) and non-routine transactions. And accounting estimates should be tested by using a substantive approach. On the other hand, for non-routine transactions such as negotiation on large vendor contracts, internal auditors are recommended to examine the contracts, review the amount ordered, and open items by comparing with the industry standards. In addition, accounting estimates about the valuation of inventory and depreciation should be tested. In this respect, internal auditors should review all related calculations by management for recoverability and usage, review estimates of useful life versus actual lives. Especially, it is important to check the estimated amount of depreciation using the prior year's amount, additions, and deletions

Source: Adapted by the Authors from Prager Sealy, KPMG, and Attain (2010); APQC (2014); IIA (2017) and Kandeh and Alsahli (2020).

Table 2.12 Endowment management process

Endowment Management process description	Endowment management is the key to providing the HE institution's long-term monetary needs. In this respect, it is important to establish a system for investments management to obtain the highest yields consistent with the HE institution's objectives. In this process, the performance of the portfolio managers is the key to success and full compliance with laws is essential
Process objectives	1. Gaining a high level of return on endowment funds X% 2. Getting transaction fees which are competitive to HE sectors 3. Minimizing the spending rate for the endowment fund to X% 4. Full conformity with donor restrictions and laws 5. Making investments based on the official investment policy guidelines of the HE institution 6. Increasing the total return of the endowment fund by $X
Evaluation factors linked to process objectives	• Distributing the funds to investments with the highest return to support spending rates of the HE institution (**3**) • Cutting costs allocated for the administration of the portfolio (**2**) • Establishing a database for compliance with laws and donor restrictions (**4**) • Controlling the funds invested by portfolio managers to ensure that all investment decisions are made in accordance with the HE institution's policies (**1, 5, 6**)
KPIs linked to process	• The ratio of return on investment portfolio • The ratio of costs incurred by portfolio management (manager fees) • The processing cost of transactions (trustees, managers, and brokers) • The total amount of spending (trend analysis) • The ratio of endowment spending as a percentage of educational and general and mandatory transfers • Measuring the relevant market value of investments over carrying value • The KPI related to each portfolio manager • The ratio of the allocation of the portfolio by investment type • Benchmarking the size of endowments in the HE sectors
KRIs linked to process objectives	a) Due to volatile economic conditions, there may be low rates of return in financial markets (**1**) b) There may be refunding of invested funds due to noncompliance with agreements and laws (**4**) c) There may be a change in the relevant discount rate or inflation rate which reduce the value of certain real estate investment portfolios (**1, 3**) d) There may be an increase in costs due to conflict of interests (**2**) e) There may be a low-performance level due to ineffective fund managers (**1**)

(Continued)

Table 2.12 (Continued) Endowment management process

	f) There may be a lack of endowment funds or other funds designated for long-term investment in the HE sectors (**6**) g) There may be a lack of surveillance due to the Board's approach who is not properly monitoring the risk of investments (derivation) (**5**)
KCIs linked to KRIs	• Making comparisons regarding the projected total return to actual return, the correlation between projections and investment class and maturities (**a, e, g**) • Establishing a control and review mechanism for confirming the rates of return and portfolio manager performance (**a, e**) • Creating a spending rate policy to cover the process, and periodic reports on donor restrictions and compliance and internal audit reports (**b, f**) • Establishing an investment policy statement, providing oversight by management or board of trustees, and generating a conflict of interest policy (**d, g**) • Diversification of investment portfolio (**c**)
Audit objectives	• Internal auditors should control and confirm that all the purchases, sales, gains, and losses are recorded properly (COMPLETENESS AND ACCURACY) • It is important to control that all interest and dividend income is recorded, and premium/discount is amortized (COMPLETENESS AND ACCURACY) • Internal auditors should control the existence of all investments and confirm that they are all owned by the HE institution (EXISTENCE AND OWNERSHIP) • Internal auditors should check the valuation methods, write-downs for impairment to confirm that they are appropriate and consistently applied (VALUATION, PRESENTATION, AND DISCLOSURE) • Internal auditors should review the off-balance-sheet financial instrument transactions to confirm that they are identified and are recorded in accordance with IFRS (COMPLETENESS AND ACCURACY) • It is important to review and confirm the existence of off-balance sheet financial instruments and their ownership by the HE institution (EXISTENCE AND OWNERSHIP) • Internal auditors should review the process to confirm that all financial instruments are properly disclosed (PRESENTATION AND DISCLOSURE)

(*Continued*)

Table 2.12 (Continued) Endowment management process

Audit test procedures	It is a fact that there are many HE institutions with a huge amount of endowments and investment portfolios that are carefully controlled by the Management Board. It should be noted that the transaction types related to this process are mostly routine and adequately assessed by CA/CM mechanisms established by the HE institution. Substantive procedures that should be considered by the internal auditors include trend analysis on portfolio performance; allocation of investments and comparison to peer group and investment type indices; and review of spending policy of the HE institution. There may be some routine transactions such as incurring and allocating salary and fringe benefit costs in this process. It is recommended for internal auditors to apply the following tests of controls: Determining the previously defined KCIs significantly reduces the risk of misstatement and test them; performing data analytics using applicable KPIs such as yield analysis of investments compared to prior years; comparison of portfolio allocation to previous years; comparison of yields to investment indices and peer institutions; compare amount used for spending to approved spending rate; compare year-end balances to the prior year and perform two to three-year trend analysis. Internal auditors should investigate results, as necessary. On the other hand, internal auditors should apply a substantive test approached for the non-routine transactions and accounting estimates. One example of frequently faced cases of non-routine transactions is that transfers between portfolio managers. In this case, internal auditors should review the transfer of assets and determine if inappropriate gains/losses were recognized. Another example of non-routine transactions is the purchase of complex investment instruments. Internal auditors should review the related agreements regarding the purchase investments, control the accuracy of accounting for securities, determine the relevance of income recognition, confirm the recoverability and valuation of the investment. It should be noted that internal auditors should apply accounting estimates to determine accrued interest receivable, accretion of premium/amortization of discount, check the calculations for accuracy, and consider as part of yield analysis. In addition, internal auditors should make a comparison for the current market value to the carrying value regarding the valuation of investments

Source: Adapted by the Authors from Prager Sealy, KPMG, and Attain (2010); APQC (2014); IIA (2017) and Kandeh and Alsahli (2020).

Table 2.13 Financial management process

Financial Management process description	The financial management process is the key for the HE institution's monetary needs considering the oversees preparation of the HE institution's operating budgets and capital budgets. In this respect, the management of the short-term and long-term cash investments to gain the highest yields is important. The financial management process also involves monitoring compliance with regulations and taxes.
Process objectives	1. Decreasing the operating budgets of the HE institution by X% compared to the prior year's levels 2. Establishing management of short-term and long-term investments to achieve yield an average return of X% or higher 3. Controlling the refinance needs related to the existing debt so that the average rate is X% 4. Decreasing the credit line for borrowings to $X per month and Y days outstanding 5. Keeping the tax-exempt status and complying with the regulations
Evaluation factors linked to process objectives	• Establishment of high-quality and sustainable relationships with lenders and other sources of funds (**4**) • Investing cash to maximize return and minimizing the idle cash as much as possible based on financial instruments (**2**) • Monitoring closely negative budget variances for taking necessary actions (**1**) • Management and reduction of debt carrying costs (**3**) • Controlling the tax exposure to prevent the loss of exempt status (**5**)
KPIs linked to process	• Defining the cash needs periodically (for the next month, quarter, year) • Measuring and preparing the budget analysis to capture the negative budget variances • The ratio of interest expense to total expense • The ratio of cash on hand to short-term borrowing • Yields on short-term investments • The ratio of expendable/nonexpendable fund balances to total expenditures and mandatory transfers • The average rates obtained on borrowings
KRIs linked to process objectives	a) There may be a decrease in the interest rates (**2**) b) There may be a lack of lenders in the markets (**3**) c) There may be a rise in the cost of finance and oversight considering the lending agencies (**1, 3**) d) There may be some delays/failures of customers to pay their bills on time (**2, 4**) e) Losing the tax-exempt status and there may be lack of compliance to the regulations (**5**) f) There may be some penalties and additional taxes for unrelated business income (**5**) g) There may be a lack of effective oversight over the budgeted to actual amounts (**1**)

(*Continued*)

Table 2.13 (*Continued*) Financial management process

KCIs linked to KRIs	• Making comparisons for the budgeted expenditures to actual expenditures regarding the appropriate and properly approved adjustments (**g**) • Controlling the status of the debt covenants and related financial ratios annually (**c**) • Benchmarking the cost of debt to analyze the HE sectors norms and general business norms (**b, c**) • Auditing the financial statements (**c**) • Controlling the returns and related maturities on short-term and long-term investments accordingly (**a**) • Determining the cash flow needs on a timely basis (**d**) • Monitoring tax issues to prevent any inconvenience (**e, f**)
Audit objectives	• Internal auditors should confirm the existence of the Cash to confirm the availability to meet the obligations of the HE institution, and the cut-off is proper (EXISTENCE, ACCURACY, COMPLETENESS, VALUATION, PRESENTATION, AND DISCLOSURE) • It is important to determine that all debt is accurately recorded (EXISTENCE, OWNERSHIP, COMPLETENESS, AND ACCURACY) • It is recommended to check the interest expense to confirm that it is recorded properly (COMPLETENESS AND ACCURACY) • Internal auditors should review the process to give reasonable assurance that the accounting principles are applied appropriately (PRESENTATION AND DISCLOSURE) • Internal auditors should investigate the interfund borrowings to control that they are all approved, proper, and will be repaid (VALUATION, PRESENTATION, ACCURACY, AND DISCLOSURE)

(*Continued*)

Table 2.13 (Continued) Financial management process

Audit test procedures	The financial management process is essential for the HE institution to create a sustainable funding strategy concerning the short-term investments, cash, and cash flow as well as obtaining necessary financing in the long term. This process is also key to the maintenance of the HE institution's tax-exempt status. The financial management process involves various financial transactions such as payroll, receivables, and investments about the other core business or resource management processes, i.e., human resources, education, and endowment are also closely related to this process. It is a fact that cash and short-term investments are mostly routine. While transactions concerning debt occur more infrequently. Internal auditors should determine the routine transactions by applying the following test of controls: Determining which of the previously defined KCIs significantly reduce the risk of material misstatement and testing them. Internal auditors should use the substantive procedures to conduct performance analytics by using applicable KPIs; confirmation of debt, cash, and short-term investment accounts; analyzing cash and investment balances and investment yields compared to the prior years and economic factors; determining interest costs compared to the prior years and stated rates; comparing the asset and debt balances with third-party information. On the other hand, there may be some risky non-routine transactions such as "Issuance of Debt", "Debt Repayment", "Interfund Borrowings", and "Valuation of Complex Investments". Internal auditors should check these non-routine transactions in the following ways: • *Debt*: Controlling the new debt agreements/financing arrangements; controlling the compliance with covenants and repayment schedule; confirming the current debt level; determining the interest expense calculations. • *Interfund borrowings*: Checking the existence of the board of trustee or management approval; controlling the HE institution's ability to repay the debt. • *Valuation of investments*: Comparing the current market value of investments to the carrying value

Source: Adapted by the Authors from Prager Sealy, KPMG, and Attain (2010); APQC (2014); IIA (2017) and Kandeh and Alsahli (2020).

Table 2.14 IT management process

IT Management process description	This process is essential and gaining more importance for the HE institutions in various aspects in the future. The IT management process is used for planning, implementing, maintaining, and operating systems for collecting, processing, and distributing relevant information. In this respect, this process involves the technology and digitalization needs of the core business and resource management processes. The IT management process includes two major sub-processes namely, administrative, and academic technology. It is a fact that this process will become central to the HE institutions in the next several years
Process objectives	1. Increasing the efficiency by improving the reports delivered within deadlines by X% from the prior year 2. Decreasing the days to finish the requested system changes by X days 3. Improving online inquiry response time by X% 4. Decreasing the operational costs of the IT management process by $X 5. Improving the transaction processing costs by $X 6. Increasing the number of students using computers by X% 7. Increasing the usage of technology by faculty by X% 8. Increasing the reliability of data belonging to the HE institution 9. Decreasing inappropriate access to information and improving the IT security
Evaluation factors linked to process objectives	• Establishing IT systems that provide accurate, timely, and reliable information for the HE institution's digital needs and expectations (**1, 3, 8**) • Including related users in the technology acquisition process to achieve sustainability (**6, 7**) • Developing digital business processes simultaneously so that IT systems optimally support related processes and hence, the process owners will perform their roles and responsibilities by optimum use of systems (**1, 2, 3, 8**) • Generating a cost-effective purchasing culture of technology (**4**) • Establishing systems at the lowest possible cost for the HE institution (**5**) • Providing modern IT systems for faculty and student use (Internet, video conferencing, etc.) (**6, 7**) • Developing relevant policies and procedures regarding access control/passwords protection systems (**9**)
KPIs linked to process	• Time span between the transaction occurring time to the information processed and reported time • Response time considering the online queries • The share of information management costs compared to the total costs • The percentage of information management staff to total staff • Return on investment considering the precise IT projects • The share of IT operations costs compared to the outsourced services

(Continued)

Table 2.14 (Continued) IT management process

	• Satisfaction surveys for the internal IT users in the HE institution • Total number of classes implementing technology-based resources • The share of virtual participation in classroom education
KRIs linked to process objectives	a) There may be a lack of competent staff due to inadequate training of use of technology (**1, 2**) b) There may be some difficulty in obtaining IT systems due to higher prices (**4**) c) There may be some Natural/man-made disaster/catastrophes (**1, 2, 3, 5**) d) There may be some erroneous/inaccurate or untimely decisions made by the HE institution based on lack of system reliability, integrity responsiveness, and usefulness of information (**1, 8**) e) There may be cases where the needs of users (administrators, faculty, and students) are not met. This may be due to ineffective, costly technology, or lack of knowledge (**6, 7**) f) There may be a lack of ability to use new technology within the HE institution (**6, 7**) g) Theft of information/data (**9**)
KCIs linked to KRIs	• Applying the system development life cycle methodology by including the users (**e**) • Reviewing system performance statistics regularly (**d**) • Creating an IT steering committee (including administrators and faculty, where appropriate) who will oversee monitoring the IT costs, utilization, and sufficiency of IT systems (**b, d**) • Comparing IT costs to the HE sectors norms and external providers (**b**) • Establishing the IT security policies and procedures (**g**) • Preparing disaster recovery plan (**c**) • Establishing the backup, archival, and record retention policies (**c**) • Providing effective training for IT staff and key personnel (**a**) • Providing effective training/workshops/demonstrations for all users (**f**)
Audit objectives	The audit approach for this process should be based on the procedures performed in understanding the control environment and tests of controls over general IT controls and specific IT applications in the HE institution. The major transactions incurred in this process are usually related to payroll costs of staff assigned to the information management resource process, materials and supplies used in the process, and equipment purchased and depreciated. In this respect, it should be noted that such transactions must be tested in the other support processes such as human resources and materials management
Audit test procedures	This part will be explained in further chapters in more detail

Source: Adapted by the Authors from Prager Sealy, KPMG, and Attain (2010; APQC (2014); IIA (2017) and Kandeh and Alsahli (2020).

NOTES

1 Prosumer is defined as "a prospective consumer involved in the design, manufacture, or development of a product or service".
2 API Gateway is better for management because it provides an open interface for external users. In this way, it is possible to isolate the different components of Information Systems and to increase the interactions across different units. API is also considered as a base for innovation since various new technologies have emerged from open-source implementations.
3 Changing from analog to digital form. Digitization involves organizing information and digitizing information. That is the first stage in the digital maturity level.
4 Using digital technologies to transform the institutional operations. This stage involves streamlining the processes and automation. Digitalization is the second stage in the digital maturity level.
5 Transforming the whole institution and digital transformation is the third stage in the digital maturity level.
6 It should be noted that the strategic management process does not generate a remarkable number of financial transactions. The primary transactions are related to the cost analysis based on the salary and fringe benefit cost of senior management, fees paid to trustees, traveling expenses, and other miscellaneous items. For this reason, this process does not generate any entries to financial statement accounts or off-balance sheet items.
7 On the other hand, application fees and deposits for admissions are not that significant. The deposits are generally non-refundable. They are recognized as revenue upon receipt. Since it is not that significant to the current financial statements, no Potential Audit Objectives are required. It is convenient to apply CA/CM approach for Routine Transactions – Accepting Deposits.
8 In the HE sectors, tuition and fee revenue is considered to be generally the most important revenue component. It should be noted that most of the transactions in this process are routine transactions such as "Registration of students", "Granting of financial aid", "Billing and collecting tuition, housing, and other fees". On the other hand, there is a need for an accounting estimate for calculating the allowance for doubtful student accounts and bad debt expenses.

REFERENCES

Acemoglu, D., Simon, J., James, R. (2005). Institutions as a fundamental cause of long-run growth. In Philippe Aghion & Steven N. Durlauf (Eds.), *Handbook of economic growth 1* (pp. 385–472). Amsterdam: Elsevier.

Aghion, P., Christopher, H., Peter, H., John, V. (2001). Competition, imitation and growth with step-by-step innovation. *Review of Economic Studies* 68(3), 467–492.

Amaral, A. (2014). Where are quality frontiers moving to? Quality assurance in higher education. Chapter 2. In M. J. Rosa & A. Amaral (Eds.), *Contemporary debates*. New York: Palgrave. ISBN:978-1-137-37463-9 (eBook).

American Productivity & Quality Center – APQC (2014). APQC Process Classification Framework (PCF) – Education – PDF Version 3.0.1E.

Andjelkovic, M. (2015). What is the digital economy and why does it matter? Background paper for the World Development Report 2016, World Bank, Washington, DC.
Apraxine, D., Stylianou, E. (2017). Business intelligence in a higher educational institution: The case of University of Nicosia. In *Proceedings of the 2017 IEEE global engineering education conference (EDUCON)*, Athens.
Arcuria, P. (2015). Applying academic analytics developing a process for utilizing bayesian networks to predict stopping out among community college students. PhD Thesis. Arizona State University, Arizona.
Arnold, K. E., Pistilli, M. D. (2012). Course signals at Purdue: Using learning analytics to increase student success. In *Proceedings of the 2nd international conference on learning analytics and knowledge*, Vancouver.
Arroway, P., Morgan, G., O'Keefe, M., Yanosky, R. (2016). Learning analytics in higher education. Accessed from: https://library.educause.edu/resources/2016/2/learning-analytics-in-highereducation
Aziz, A. A., Idris, W. M. R. W., Hassan, H., Jusoh, J. A. (2012). Intelligent system for personalizing students' academic behaviors – A conceptual framework. *International Journal of New Computer Architectures and Their Applications*, 2(1), 138–154.
Bozkus Kahyaoglu, S. (2019). An analysis on the implementation of new approaches and techniques in the auditing of business processes based on blockchain technologies. In B. Darici & F. Ayhan (Eds.), *Cryptocurrencies in all aspects*. New York: Peterlang Publications. ISBN:978-3-631-78387-0.
Bresnahan, T., Erik, B., Lorin, M. H. (2002). Information technology, workplace organization, and the demand for skilled labor: Firm-level evidence. *Quarterly Journal of Economics* 117(1), 339–376.
Brooks, D. C., McCormack, M. (2020). Driving digital transformation in higher education. ECAR research report. Louisville, CO: ECAR, June 2020. *EDUCAUSE*.
Bührig, J., Schoormann, T., Knackstedt, R. (2018). Business process management in German Institutions of higher education: The case of Jade University of Applied Science. In J. vom Brocke, & J. Mendling (Eds.), *Business process management cases, management for professionals*. Switzerland. DOI: 10.1007/978-3-319-58307-5_31
Capitani, G. (2018). CIO's: Drivers or followers of digital transformation? In G. Bongiorno et al. (Eds.), *CIOs and the digital transformation*. Switzerland: Springer. DOI: 10.1007/978-3-319-31026-8_9
Child, J. (1972). Organizational structure and strategies of control: A replication of the Aston study. *Administrative Science Quarterly*, 17(2), 63–177.
Child, J. (1973). Strategies of control and organizational behavior. *Administrative Science Quarterly*, 18(1), 1–17.
Christensen, C. M., Eyring, H. (2011). *The innovative university: Changing the DNA of higher education from the inside out*. San Francisco: Jossey-Bass.
Christensen, C. M., Horn, M. B., Caldera, L., Soares, L. (2011). *Disrupting college: How disruptive innovation can delivery quality and affordability to postsecondary education*. San Mateo, CA: Center for American Progress and Innosight Institute. Retrieved from http://www.americanprogress.org/issues/labor/report/2011/02/08/9034/disrupting-college/
Clarke, J., Nelson, K., Stoodley, I. (2013). *The place of higher education institutions in assessing student engagement, success and retention: A maturity model to guide*

practice. In S. Frielick, N. Buissink-Smith, P. Wyse, J. Billot, J. Hallas & E. Whitehead (Eds.), *Proceedings of research and development in higher education conference: The place of learning and teaching* (vol. 36, pp. 91–101). Auckland, New Zealand, 1–4 July 2013.

Commander, S., Rupert, H., Naercio, M. -F. (2011). ICT and productivity in developing countries: New firm-level evidence from Brazil and India. *Review of Economics and Statistics* 93(2), 528–541.

COSO. (2019). *Managing cyber risk in a digital age*. https://www.coso.org/Documents/COSO-Deloitte-Managing-Cyber-Risk-in-a-Digital-Age.pdf

Daniel, B. (2014). Big Data and analytics in higher education: Opportunities and challenges. *British Journal of Educational Technology*, 46(5), 1–17. DOI: 10.1111/bjet.12230

Grajeck, S. (2016). *The digitization of higher education: Charting the course*. https://er.educause.edu/articles/2016/12/the-digitization-of-higher-education-charting-the-course

Green Book (2017). Enhancing university board governance and effectiveness university. *Transformation Programme, Green Book*. Malaysia: Ministry of Higher Education.

Hedman, J., Kalling, T. (2003). The business model concept: Theoretical underpinnings and empirical illustrations. *European Journal of Information Systems*, 12, 49–59. DOI: 10.1057/palgrave.ejis.3000446

Hess, T., Matt, C., Benlian, A., Wiesböck, F. (2016). Options for formulating a digital transformation strategy. *MIS Quarterly Executive*, 15, 151–173.

Hinssen, P. (2010). *The new normal explore the limits of the digital world*. Belgium: Mac Media NV. ISBN:978-9-081-32425-0.

Institute of Internal Auditors – IIA (2017). International standards for the professional practice of internal auditing (standards). IPPF-Standards-2017.pdf (theiia.org)

International Association of Universities (IAU) (2019). Higher education in the digital era. The current state of transformation around the World. https://iau-aiu.net/Higher-Education-in-the-Digital-Era-The-Current-State-of-Transformation-Around

Ireland, R. D., Webb, J. W. (2007). Strategic entrepreneurship: Creating competitive advantage through streams of innovation. *Business Horizons*, *50*, 49–59.

Johnson, M. W., Christensen, C. M., Kagermann, H. (2008). Reinventing your business model. *Harvard Business Review*, 86(12), 51–59.

Kandeh, H., Alsahli, M. (2020). Effect of Big Data Analytics on Audit An exploratory qualitative study of data analytics on auditors' skills and competence, perception of professional judgment, audit efficiency and audit quality. UMEA University, Department of Business Administration Master's Program in Accounting Master's Thesis in Business Administration III, 30 Credits, Spring 2020 Supervisor: Tobias Svanström.

Kettunen, J. (2008). A conceptual framework to help evaluate the quality of institutional performance. *Quality Assurance in Education*, 16(4), 322–332.

Kettunen, J. (2010). Strategy process in higher education. *Journal of Institutional Research*, 15(1), 16–27.

Kettunen, J. (2015). Integrated management approaches in higher education. *US-China Education Review A*, 5(11), 707–717. DOI: 10.17265/2161-623X/2015.11.001. https://www.researchgate.net/publication/310813387_Integrated_Management_Approaches_in_Higher_Education

McCaferry, P. (2019). *The higher education manager's handbook effective leadership and management in universities and colleges*, 3rd ed. London: Routledge. ISBN:978-1-351-24974-4 (ebk).

McKinsey Global Institute (2019). Twenty-five years of digitization: Ten insights into how to play it right. In Prepared for the digital enterprise show 21–23 May, Madrid May 2019. *Briefing note*. https://www.mckinsey.com/~/media/mckinsey/business%20functions/mckinsey%20digital/our%20insights/twenty-five%20years%20of%20digitization%20ten%20insights%20into%20how%20to%20play%20it%20right/mgi-briefing-note-twenty-five-years-of-digitization-may-2019.ashx

McLachlan, G., Ritchie, M., Armstrong, A. (2017). Digital transformation at the University of Edinburgh. Presentation. https://www.slideshare.net/MarkRitchie2/digital-transformation-at-the-university-of-edinburgh/1

Messina, M. (2018). Designing the new digital innovation environment. In G. Bongiorno et al. (Eds.), *CIOs and the digital transformation*. Switzerland: Springer. DOI: 10.1007/978-3-319-31026-8_9

Pandey, U. C., Kumar, C. (2020). Enhancing the university outreach through collaborations and sharing of resources. Chapter 15. In U. M. de M. Azeiteiro & J. P. Davim (Eds.), *Higher education and sustainability*. Boca Raton: Taylor and Francis. ISBN:978-1-138-55653-9.

Prager, Sealy & Co., LLC, KPMG LLP, Attain LLC. (2010). *Strategic financial analysis for higher education identifying, measuring & reporting financial risks*, 7th ed. USA: KPMG.

Pusser, B., Slaughter, S., Thomas, S. L. (2006). Playing the board game: An empirical analysis of university trustee and corporate board interlocks. *The Journal of Higher Education*, 77(5), 747–775.

Rajterič, H. I. (2010). Overview of business intelligence maturity models. *Management: Journal of Contemporary Management Issues*, 15(1), 47–67. Erişim adresi: https://hrcak.srce.hr/53606

Rubin, B. (2013). University business models and online practices: A third way. *Online Journal of Distance Learning Administration*, 15(1):17 pages. http://www.westga.edu/~distance/ojdla/spring161/rubin.pdf

Sambamurthy, V., Zmud, R. W. (2017). *Guiding the digital transformation of organizations*, 2nd ed. Legerity Digital Press LLC. ISBN:978-0-9995347-0-0.

Stigler, G. J. (1961). The economics of information. *Journal of Political Economy* 69: 213–225.

Stolterman, E., Fors, A. (2004) Information technology and the good life. In B. Kaplan, D.P. Truex, D. Wastell, A.T. Wood-Harper, J.I. DeGross (Eds.), *Information Systems Research. IFIP International Federation for Information Processing*, vol. 143. 687 692. Boston, MA: Springer.

Tremblay, K., Lalancette, D., Roseveare, D. (2012). *Assessment of higher education learning outcomes feasibility study report volume 1 – Design and implementation*. Paris, France: OECD Publications.

Uhl, A., Gollenia, L. A. (2014). Digital enterprise transformation. A business-driven approach to leveraging innovative IT. ISBN:9780367670054.

Vasarhelyi, M. A., Kuenkaikaew, S. (2010). *Continuous auditing and continuous control monitoring: Case studies from leading organizations*. USA: Rutgers Business School, Rutgers Accounting Research Center.

World Bank (2016). Digital dividend. World development report. International bank for reconstruction and development/The World Bank. e-ISBN:978-1-4648-0672-8. DOI: 10.1596/978-1-4648-0671-1

Chapter 3

Lessons for a CAE

3.1 BALANCING DIGITAL EXPECTATIONS AND POSSIBILITIES FOR TRANSFORMATIVE CHANGE WITH IIA STANDARDS

It is a fact that the digital transformation process includes various digital technologies, such as 5G, artificial intelligence, deep learning, big data analysis, and blockchain (OECD 2019; EU 2020a). In this context, these technologies generate a digital ecosystem that leads to changes in the economic and social lives of people. On the other hand, digital innovation transforms the ways people communicate, discovered and produced by the impact of digital transformation. It produces chances for new markets and business models to establish the HE sectors considering new products or services. It should be noted that digital transformation directly impacts the efficiency of HE institutions. Moreover, HE institutions are expected to trigger and guide the digital transformation process. This is because HE institutions are the drivers of digital innovation, including the provision of the skills required to lead this dynamic change pattern in the digital ecosystem.

The education system is required to adapt in order to take advantage of new technologies and tools. Furthermore, HE institutions are also required to develop strategies and actions to play an active role in the digital transformation process. In this context, the "Going Digital Toolkit"[1] has been prepared by OECD as a tool for the development of countries' "digital transformation strategies" (DTS) for achieving digital economies. The key indicators of a DTS are a strategic vision for the digital transformation, priorities, measurable targets, adequate budget, and solid monitoring and assessment.

The meaning of digital transformation for the HE institutions is that firstly, there must be new digital processes, adaptation of new digital teaching methods and tools. Secondly, there must be an online education system based on helping students in achieving the skills and competencies essential to respond and satisfy the needs and expectations in digitalized societies and economies.

DOI: 10.1201/9781003093008-3

HE institutions should set up adequate policies and procedures to support the public policies by helping companies, in particular small- and medium-sized enterprises (SMEs) to adopt emerging technology and obtain suitable digital skills for their staff (OECD 2019). It is not only the SMEs but also some start-ups and spin-offs that can benefit from partnerships with HE institutions. In this way, these organizations can acquire the initial know-how, equipment, and funding to utilize new technologies and enhance new products and services about the new research results in the digital ecosystem.

The digital transformation process becomes an essential part of actively supporting innovation in all HE institutions' missions and strategic objectives and missions in all its dimensions. That brings a dual perspective for the HE institutions namely, "inter-institutional" and "intra-institutional" behaviors (Cantú-Ortiz and Fangmeyer Jr. 2018). In this context, "intra-institutional" behaviors mean the one internal to the organization with the digital transformation of HE institutions themselves. That requires a new mindset by considering the challenges and opportunities brought by digitalization and new digital business processes assisting all the students, staff, and researchers.

"Inter-institutional" behaviors relate to the role of HE institutions that can be described as fostering digital innovation and supporting a wider ecosystem formed by all firms, institutions, and stakeholders, respectively. In this way, they will be able to jointly pursue the effort of innovation and growth through the means of innovative and new digital technologies.

For HE institutions to meet both their internal needs and external needs toward the society in the digital transformation process, sound strategic planning is required. Especially, among the obstacles to digital transformation, the lack of trust in technology and the virtual environment is decisive. To overcome this, internal control, internal audit, compliance, and risk management functions must be efficient and well-established (Kahyaoglu Bozkus and Caliyurt 2018). In this respect, there is a need for a risk management framework to develop policies to increase trust, including assessing and managing risks related to digital technologies, big data, and workflows. Internal auditors play a key role in this framework to ensure that digital security risk is a strategic priority for students, staff, faculty, and the HE institution as a whole in particular. In this way, it will not be treated as a technical question of the risk management team, and that everyone will have the responsibility for managing digital risk (Dai 2017; COSO 2019).

The point to note here is that neither the internal audit department nor the other related departments can be included in every project in the HE institution, or it is not possible. However, PwC (2019) argues in the study of "State of the Internal Audit Profession" that if the internal audit team is included in the digital transformation strategy from the very beginning, they can provide advisory services in line with this goal and strategy. Thus, they can contribute to the early detection of risks that may arise in the digital transformation process.

According to the study conducted by PwC (2019), it is recommended that internal auditors make necessary updates on six fundamental issues to make the risk management process of their institutions effective. These are as follows:

Involvement in the digital plan of the HE institution: First, they need to be actively involved in the digital transformation strategy of the HE institution. Thus, it will be possible for the internal audit team to understand the process well, to present suggestions that may contribute to the digital transformation of the university, and to convey the control deficiencies observed to the management (Kovalenko et al. 2019). Otherwise, the internal audit mechanism cannot function. Therefore, the chief audit executive (CAE) needs to plan the audit to balance between IIA standards and the needs required by the digital transformation process.

Investment into the talent and digital skills of the audit team: It should be noted that internal auditors are not required to have a deep technical knowledge to begin engaging in the HE institution's digital initiatives. Rather, internal auditors are expected to add value to decision-makers with their risk-based approach that provides a thorough understanding of risk and process. Internal auditors could associate the risks across multiple digital ambitions throughout the HE institution. On the other hand, to achieve this, internal auditors should improve their digital skills and level of digital acumen as the HE institution becomes increasingly digital. For instance, internal auditors are required to have a deeper understanding of big data which is critical to gain foresight because big data is in the center of all learning and research environments with digitalization (Esteves et al. 2019). An initial point to start for internal auditors can be mentioned as, even though they do not need to be data scientists, they do have to comprehend the sources of data. This is important to determine the data quality, to examine whether an AI algorithm is performing as planned, and to identify what kinds of foresight can be drawn from the big data.

Digital fit for emerging technologies: It is a fact that when it comes to using the new audit technologies within the internal audit function, many of them struggle to find the right fit. There are various emerging technologies such as RPA, machine learning, deep learning, and AI which provide the opportunity to modernize and automate all kinds of processes and to achieve greater assurance over internal control effectiveness. To give an idea about how it helps internal auditors, consider a HE institution where internal auditors are testing to determine whether terminated staffs' system access rights were being removed promptly. This testing procedure is mostly achieved on a manual audit process. For this, it is required to apply a lookup function from disparate data sources for each IT application, which took the audit team many hours to test a few audit samplings of the control (Anissa et al. 2019). On the other hand, if AI techniques are applied, by automating many audit stages of the test except auditors' review, testing hours will be greatly reduced. In addition, the audit fieldwork coverage will be expanded from a sampling basis to full populations, i.e., audit universe, which provides greater assurance.

Delivering the agile audit for responding to risks in real-time: It is a fact that internal audit executives commonly agree that annual plans and annual assessments are "antiquated". In such a speedy and digitally transforming business environment in the HE sectors, they need more frequent and more fluid audit cycles than ever before. In this respect, most of the internal audit functions revise their risk assessments and audit plans more frequently than they used to. As HE sectors are speedily moving to online education systems, internal audit functions should be applying the "agile auditing" strategy considering the planning, testing, and validating in "sprint cycles" rather than waiting until projects finish before they present their audit recommendations. In this regard, PwC (2012) interviewed Melvin Flowers as part of a research study. Melvin Flowers who is the Corporate Vice President in Microsoft Corporation and responsible for Internal Audit, states that they do not need to make a long-term internal audit plan due to the rapid change environment they are in, and instead, they have to implement a more dynamic and agile audit strategy in which they update their annual audit plan every three months.

Collaboration with the digital leadership: It is a fact that the digital transformation in the HE sectors requires a move in the maturity level for internal audit. Because the baseline contribution from internal auditors expected by the stakeholders is rising every day. Therefore, the potential value internal audit can deliver is expanding rapidly in the HE sectors. Internal audit functions should advance their role to consider that their organizations are managing risk effectively in a digital business environment (Luckin et al. 2016). Hence, internal audit functions that are digitally fit are well-positioned to collaborate with the digital leadership on their way to achieving the strategic objectives of the HE institution.

Alignment to provide a consolidated view of risks: According to a joint work of Risk Management Society (RIMS) and Institute of Internal Auditors (IIA) (2012), as an organization's risk maturity level increases, risk management will become more embedded in the business processes. In this respect, the role of internal auditors in championing the ERM framework may reduce. In other words, if an organization establishes a risk management function, internal auditing is more likely to deliver value by concentrating on its assurance role than by undertaking advisory activities (Kahyaoglu Bozkus and Aksoy 2013). It is expected that internal audit and risk management functions cooperate and increase efficiency in the context of integrated risk management (Albrecht 2008;

Deloitte 2012; COSO 2019). Thus, the involvement of internal auditors in the early detection and management of risks is of critical importance to address the risk universe with a holistic approach.

In the digital transformation process, important decisions are made, and critical steps are taken in all universities. While the flexible and remote working environment based on digital business processes provides comfort to all personnel, it also brings risks in the virtual environment. In such an environment, the most fundamental problem is whether internal audit capabilities are sufficient to detect risks early that arise in the digital transformation process (Deloitte 2012). It is important to carry out audit field works with a modern approach by providing the agility and flexibility required by a digital transformation without compromising internal audit professional practice standards (Chartered Institute of Internal Auditors 2020). In other words, internal audit teams should adapt their audit methodologies to assure the digital workplace of the HE institutions.

3.2 RETHINKING AUDIT CULTURE: WILL IT BE A DIGITAL PAIN OR DIGITAL POWER?

While higher education institutions lead the rapidly developing digital transformation that is spreading irreversibly all around the world, they must identify their corporate risks early and take necessary precautions. We have previously stated that the common feature of HE institutions that achieve and implement digital transformation is to transform their corporate culture into digital culture. In this context, it is recommended that internal audits be functional so that the cultural transformation is carried out without deviations and following strategic goals (Ramamoorti and Siegfried 2016).

For internal audit to fulfill such a strategic role properly, it must first be ensured that it harmonizes its own audit culture. Internal audit will be a guide in shaping the targeted digital culture in the universities and ensuring its effectiveness, with its value-added recommendations. This situation requires the internal audit itself to review and restructure its business processes as a department. It should be in mind that the internal audit department is prone to early detection of needs and expectations in the context of "quality assurance review" (QAR) services and to make necessary adjustments.[2] The QAR process of the audit, which is expected to be every five years at the latest, is the basis for this. KPMG (2013) recommended that internal audits should be reviewed in the context of people, processes, and positioning (*3P approach*) within the organization (Figure 3.1).

Higher education institutions have complex institutional environments with various cultural dimensions. McDaniel et al. (2007), with reference to higher education institutions, explained that complex systems share five common characteristics: (a) the first of these is a new digital communication

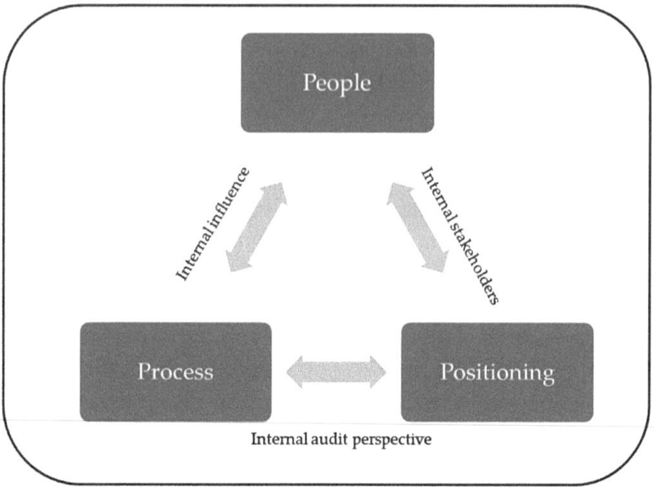

Figure 3.1 3P approach for internal auditing.

Source: KPMG (2013).

platform with a wide range of stakeholders, (b) second, there may be asymmetric, non-linear, and even stochastic connections in the digital ecosystem, (c) third, with the emergence of autonomous institutional structures, self-sufficient digital structures will emerge, (d) Fourth, new risks may arise in unpredictable new digital systems, and (e) fifth, there may be digital systems that support the development of the digital business environment.

Accordingly, developing technological infrastructure only digitally is not enough to achieve the strategic objectives of HE institutions in such complex structures. It is seen that those working in such a business environment are open to improvement in terms of understanding, using, and internalizing the technology in question.

With the increasing dependence on technology, it is essential to prioritize increasing the competencies of those who use them. Based on the ecosystem approach in digital structuring, the development of an individual HE institution is not enough. The aim is to increase the competencies of the whole system and increase the added value in this way. The point to remember is that the system is as strong as the weakest link. Therefore, detecting the weakest link means identifying digital risks early. Hence, the internal audit mechanism has an important role in revealing this.

Internal audit departments have evolved significantly over time in terms of maturity. Kahyaoglu Bozkus (2019) argues that in line with Industry 4.0, Audit 4.0 should now be considered a basic need (Figure 3.2). What is decisive here is the application of advanced technology usage level as well as the application of remote control and social distance norms, which are accepted as new-normal.

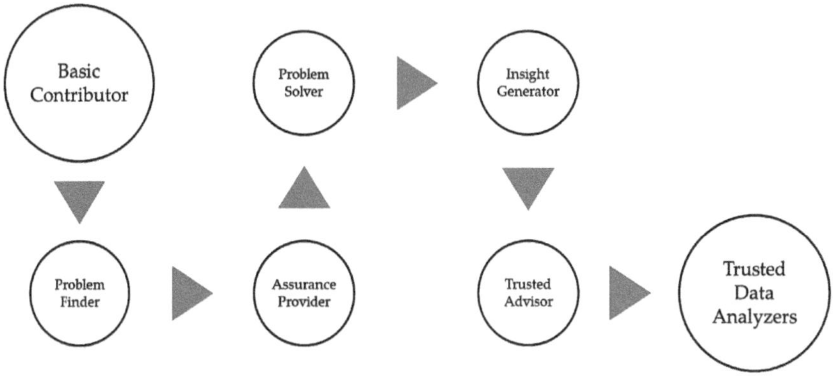

Figure 3.2 Changing roles of internal audit.

Source: Dai (2017) and Kahyaoglu Bozkus (2019).

It is a fact that internal audit is among the support service units within the organizational structure. IA prepares audit plans, determines risky areas and audit priorities, and performs field works depending on the general maturity level of the organization. In this context, internal audit needs to develop a strategy according to the maturity level of the corporate structure to add value to the organization. Hence, for the CAE to use advanced audit techniques and technologies in audit field works, it would be appropriate to make a preliminary evaluation on the following three key issues (Deloitte 2020):

1. First, the internal audit department should obtain information about the data structure, institutional maturity level, and technology use of the institution to be audited. At the same time, the technology aptitude of the internal auditors who work in the internal audit team should be taken into consideration
2. Second, legislation compliance, internal control environment, and management understanding should be examined within the processes and sub-processes of the institution
3. Third, the organization's risk appetite, risk capacity, and risk management processes should be considered. In general, strong communication, and coordination between internal audit, and enterprise-wide risk management (ERM) can support the effective use of resources by increasing the added value provided by the audit

Within the information obtained on these three basic issues, the strategy of the CAE toward digital transformation in field works is also shaped. Furthermore, the information about the maturity level of the internal audit function in the HE institution will also reveal whether the audit executed in the digital environment will turn to be a source of pain or a power source. In this context, a sample of self-assessment questionnaire set is recommended by Deloitte Risk Analytics (2020) to determine the level of corporate maturity level of internal audit. It is presented in Appendix 1. In addition, self-assessment question sets used to determine the internal audit maturity levels of strategy, people, process, data, and technology which consist of sub-criteria as basic indicators of maturity level, are given in Appendixes 2, 3, 4, and 5, respectively.

Within the framework of this self-assessment measurement, as the digital maturity level of internal audit increases, the internal audit function can generate more added value in field works and thus, this situation has become the "digital power" for internal audit. However, in the opposite case, "digital pain" may be encountered due to the low digital maturity level (KPMG 2020).

3.3 CRITICAL PERSPECTIVES ON THE NATURE OF DIGITAL COMPETENCY: INSTITUTIONAL AGENDAS FOR THE AUDIT COMMITTEE

As is well known today, in its complex and constantly evolving HE sectors, audit committees (ACs) can make a tremendous contribution toward sustaining a "no surprises" control environment. In the HE sectors, as in other sectors, the existence of an AC that works effectively, efficiently, and economically should also be a key feature of a strong, competitive, and effective governance culture. In this way, it should bring significant benefits to the HE institution (CUC 2008; EU 2019).

ACs must be supported by the basic building blocks recommended by IIA standards: a structure and foundation in line with the purpose and vision of the organization, a comprehensive and well-defined regulation containing duties, authorities and responsibilities, accurate understanding, and analysis of existing and emerging risks. In this context, it is necessary to continue surveillance activities with a proactive, risk-based approach to audit field works to support teamwork (EU 2019; KPMG 2020).

It is essential to realize how digital capabilities generate value in the internal audit functions of the HE sectors. To consider both digital capabilities and value in an integrated approach, Blaschke et al. (2017) propose the concept of "digital value drivers". These are also defined as the value-generating effects. They are related to the digital key elements and can be devoted to certain units of the enterprise view. Within the scope of a study carried out by Vuorikari et al. (2020) in the European Commission's Joint Research Center (JRC), the developments expected to occur in HE institutions with the effect of digitalization were transformed into scenarios. These scenarios are discussed in eight subjects based on future expectations which are shown in Figure 3.3. If these expectations come true, the ACs and CAEs will have an important role in understanding the events and preparing them institutionally. The approach of CAEs should strongly

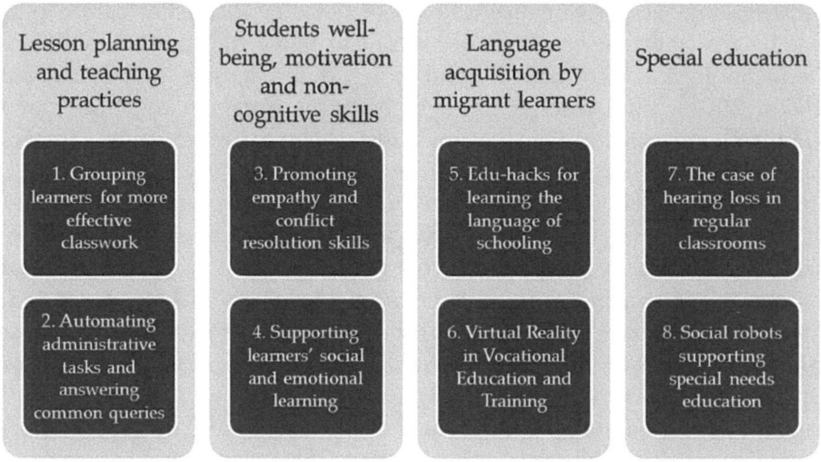

Figure 3.3 Near-future scenarios for HE institutions.

Source: Vuorikari et al. (2020).

consider educational, social, and ethical values, and should not be limited to technical issues.

The crucial point here is to identify the near-future scenarios supported by the EU's JRC, the key issues that prevent today's university lecturers from providing quality education, and to solve such problems. New digital developments will take place in classrooms, lecture halls, education centers, and digital learning environments as key areas where new technologies can be enforced to support these educators in their scientific, and academic professions.

Considering a university that wants to adopt and implement the near-future scenarios discussed in Figure 3.3, to implement a digital business model effectively, it is important to analyze the complementarities structure among the fundamental elements of the digital transformation process (Stieglitz and Foss 2015). It is possible to face unique challenges that vary depending on the goals the HE institution aims to implement, as well as top management and leadership style. In such a transition process, CAEs have key roles and duties.

In HE sectors with a dynamic structure, it is necessary for ACs and CAEs to address the three main challenges related to the complementarity framework of the digital transformation process and to provide reasonable assurance and advisory services (Stieglitz and Foss 2015). These are as follows: First, there may be co-existence of complementary elements in the current business model with traditional characteristics involving constraints arising from inertia; second, there may be difficulties in ex ante planning; and third, the problem of maintaining consistency between business model elements considering the daily routines and habits.

Estimating the performance impacts of the digital transformation process by management becomes more difficult when the changes made are the technological infrastructure and more comprehensive (Siggelkow and Rivkin 2005). Infrastructure changes have the potential to affect many complementary factors at the same time. More comprehensive changes mean a more significant deviation from existing knowledge about the digital business model, leading HE institution process owners and the leadership to a new unexplored area. An important achievement expected from the complementarity perspective is the realization that each element must fit each other to achieve its full potential (Milgrom and Roberts 1995; Porter 1996; Brynjolfsson and Milgrom 2013). In this context, it is necessary to investigate and learn how the elements in question complement or replace each other to ensure harmony and consistency in the HE institution. However, maintaining harmony and consistency also requires balancing and integrating the relationship between these elements. This creates a clear source of tension in management and employees of HE institutions in the digital business model transformation process. The most important reason for this is the difficulty in meeting and balancing conflicting demands and expectations of stakeholders. Potential stress points that may be experienced in the process can also cause risk sources and internal control weaknesses. In that sense, the internal audit function has a very critical role in detecting these early and making the necessary recommendations for policy change.

The coordination of management and surveillance functions provides important information in the success of the digital transformation process (EU 2019; Flowers 2019). Especially, to foster institutionally, the top management should ensure that this transformation is appropriately supported and rewarded through the design of improvements in the current business, key performance indicators (KPIs), and performance measurement systems. The critical steps here are analyzing the processes, determining the appropriate data for measurement, and testing and applying the indicative areas. Here too, internal audit maintains its key position in detecting potential failures and providing feedbacks for the opportunities.

In this context, such a digital transformation process in universities carries the risk of being replaced by the destructive changes that have occurred in the digital education sector. This point highlights the importance of the internal audit's responses as an "observer", based on analysis and reviews of business processes, and the value-added recommendations associated with it (KPMG 2013; EU 2019). On the other hand, it is the responsibility of the top management to monitor the external environments and to make the necessary decisions on whether the current business model is applicable or not in the HE sectors.

Competencies are decisive for the processes of change, innovation, and transformation to achieve the strategic goals and objectives of higher education institutions. Digital talents and competencies can be defined as the skills and knowledge of individuals who work in organizational units that conduct a particular activity. Tushman and Anderson (1986) define "competency-destructive innovation" in the literature for the first time to identify the destruction caused by new technologies. These authors state that competency-destructive innovation creates a gap between previous dominant technologies that the skills and knowledge base required to practice the substantial technology shift. It should be noted that "competency-destructive" innovations bring dramatic transformational changes.

Employees who are in key positions should be trained in order to eliminate incompatibilities in competencies that are required in the digital transformation process. From this point of view, investing in human resources is expressed as the determining factor in the success of technological change and digital transformation. A failure to invest in human resources and the emergence of problems in this regard may lead to a costly, unsuccessful digital transformation process (EU 2019; PwC 2019).

To analyze this digital transformation process and "audit and technology" needs in the digital environment, the Association of Chartered Certified Accountants (ACCA) prepare a report namely "Audit and Technology". This report is joint work with the Chartered Accountants of Australia and New Zealand (CA ANZ). McGhee and Grant (2019) from ACCA and CA ANZ, respectively, have prepared this report to investigate how the control is affected by technology. They classify the digital development and competency areas of the auditors and present their needs with a systematic approach. Based on the report, the critical developments are affecting the structure, execution, scope of audit activities, and those who do the job. These critical developments can be summarized as follows: "distributed ledger technology (DLT)" based on blockchain infrastructures, "robotic process automation (RPA)", "smart contracts", "drones technology", "artificial intelligence (AI)", "machine learning (ML)", "natural language processing (NLP)", "deep learning (DL)", and finally big data analytics regarding "cloud technologies". In this context, the AC should review the audit business model and determine the need to make the necessary updates (Kahyaoglu Bozkus 2019; Kahyaoglu Bozkus et al. 2019). Although the level of audit maturity revealed in recent studies is sufficient for data analytics, it has been determined that the targeted level has not been reached in the field of artificial intelligence technologies (EU 2019; UNESCO 2019). These critical changes and technological innovations can be summarized as follows:

The rapid increase in data volume: According to the research in the literature, it has been determined that the number of transactions and data volume in enterprises has increased rapidly especially since 2016. In addition, this data size is expected to continue to increase further in the future. According to Marr (2018), it is estimated that more than 90% of data on a global scale has been generated since 2016. It is stated that there will be more and more heterogeneous data generated in the future and a significant part of these data consists of financial data (Jarke and Breiter 2019).

Business model change requirement: Depending on the need for business model change in the HE sectors, internal auditors are expected to understand what is happening in the digital transformation process and understand what type of emerging risks may arise (D'andrea 2007). It should be noted that it may be impossible to realize value-added business ideas and provide audit recommendations without positioning the business model correctly and knowing its scope and business context.

Transition to automation: Transition to automation involves the realization of routine and manual work and operations automatically with system support. Accordingly, it is expected that there will be a significant transformation in the audit universe and risk universe of internal auditors. However, it should be noted that this situation will not diminish the importance of internal auditors. On the other hand, internal auditors need to improve their soft skills for better communication, persuasion, and empathy to conduct successful field works (EU 2019).

From this point of view, it is expected that the auditors of the future will have the competence that internalizes technological developments and is effective in project management and human relations. Perhaps most importantly, it is aimed to have a high ability to adapt to change (Siggelkow and Rivkin 2005; Ramamoorti and Siegfried 2016; EU 2019).

3.4 GAINING A "SEAT AT THE TABLE" WITH DIGITAL TALENTS

With the effect of artificial intelligence and digital transformation, universities need professional management understanding and technological competencies to transform their infrastructure and business models faster. In this context, they also have to train employees in the digital sectoral areas of expertise (Marr 2018; Jarke and Breiter 2019). For example, areas that require digital competence addressed in faculty structures and curriculum can be expressed as FinTech, RegTech, MedTech, EduTech, WealthTech, PropTech, InsurTech, etc. Internal audit, internal control, and risk management infrastructure must be functional to ensure the transparency and accountability of the management approach in digital transformation processes in higher education institutions (EU 2019).

The point to be taken into consideration here is that internal audit, internal control, and risk management functions should be agile and contribute to the development of processes by providing advisory with innovative approaches. On the contrary, when its existence becomes a source of bureaucracy in the HE institution, it can lead to a departure from strategic goals. Hence, this is not a desired situation at all (Jones et al. 2012). It is necessary to attach importance to the departmental organization, recruitment, and training infrastructure activities to ensure that the audit, control, and risk management teams that provide assurance services in universities have

Technical skills and ethics (TEQ):	• The skills and abilities to conduct actions consistently based on a pre-defined standard, and at the same time sustaining the standards related to "integrity, independence and skepticism".
Intelligence (IQ):	• The capability to gain and utilize knowledge –"thinking, reasoning and solving problems".
Creative (CQ):	• The capability to apply existing knowledge into a different case, by making connections, discovering potential outcomes, and creating new ideas.
Digital (DQ):	• The level of awareness and application of current and new digital technologies, capabilities, procedures, and strategies.
Emotional (EQ):	• The capability to determine emotions both yourself and others, and using and applying them to diverse tasks, and controlling and managing them.
Vision (VQ):	• The capability to predict future trends by calculating trends, facts, and figures, and filling the gaps in knowledge by reasoning with an innovative manner.
Experience (XQ):	• The capability and skills to sense stakeholder needs and expectations and meet required outcomes and create value.

Figure 3.4 Digital talent requirements for future auditors.

Source: ACCA (2016).

digital competencies and make the necessary contributions to the HE management in this field.

ACCA (2016) presents valuable research findings and insights about the required future skills mix of internal auditors and how they may change, including the "digital workplace". These digital talent requirements for future internal auditors are defined as "Technical and ethical (TEQ)", "Intelligence (IQ)", "Creative intelligence (CQ)", "Digital (DQ)", "Emotional intelligence (EQ)", "Vision (VQ)" and, "Experience (XQ)" briefly given at Figure 3.4.

With this rapid increase in data size and transaction volume, auditors have to analyze much higher volumes of data than before. Therefore, auditors are required to be equipped with the latest technological tools available to be more prepared for the future business environment. In this way, they will gain a "seat at the table".

3.5 LEGAL VS. ETHICAL ISSUES IN THE DIGITAL ENVIRONMENT: THE CAE'S DILEMMA

Much emphasis is placed on the positive developments provided by the digitalization and digital transformation process in the HE sectors and specifically for the internal auditors. However, it should be noted that besides its benefits, there may be some ethical and moral issues that should not be overlooked. This situation creates a dilemma for CAEs. The technological developments that constitute the source of this dilemma and the evaluation of different perspectives regarding them are as follows.

It should be noted that there is a need for a digital culture shaped by a "trusted digital education ecosystem" in the future with a high-quality content management structure, user-friendly design, value-added service approaches, and secure digital platforms. It is important to maintain privacy in such a digital platform supported by high ethical standards. An ethical decision-making mechanism is needed based on all kinds of analysis and examination performed through the analysis of data and machine learning after data acquisition (Massy 2016; EU 2019).

To ensure the assurance of such a digital platform, an internal audit should closely monitor the business culture and test the existence of ethical standards, rather than focusing only on business processes. In this context, it is anticipated that "ethics audit" will gain more importance among the fieldwork and audit types expected from internal audit in the future (Ramamoorti and Siegfried 2016; EU 2019; Kahyaoglu Bozkus 2019).

One of the key issues when applying new technologies that arise in the HE institutions is related to the "ethical decision-making process" that causes ethical concerns.[3] Although it is accepted that artificial intelligence will have more widespread use in the future, it is estimated that there may be difficulties in compliance with ethical and legal procedures during use. The position of internal audit is crucial in the process of identifying these challenges and potential risk areas and developing a response to them. EU (2020b) state the importance of ethical and legal consideration regarding AI implementations. There is a need for balancing between "human autonomy" and machines, the "datafication of education", and pedagogical models to prevent any inconvenience in the HE sectors.

By demonstrating a fundamental need across the globe, UNESCO (2019) has suggested that AI can be a useful tool to the achievement of "Sustainable Development Goal 4". This is possible by AI since it can be used to provide inclusive and equitable education. In addition, it enables promoting lifelong learning opportunities for all based on opportunities for personalized learning at scale. Effective internal control and audit gain importance in this process. It should be noted that considering the use of AI in higher education, an ethical approach is needed not only for protecting against negative applications of this technology but also for avoiding the insufficient use of resources that could impact learners positively (OECD 2020a, 2020b). To ensure the continuity of ethical artificial intelligence practices, the internal audit function should identify possible risks and advise senior management on the establishment of necessary control activities accordingly (IIA 2017).

The issues that might be the main risk source for ethical AI applications can be summarized as follows:

1. *Human judgment and supervision to achieve ethical AI*: To prevent AI from causing possible risks, it should be open to supervision and surveillance by humans, even though it has the feature of self-learning and application. Considering that artificial intelligence systems are

spreading rapidly; as a result, if there is a decrease in independent and introverted thinking and defects in the development of high-level thinking skills and self-regulation, which are peculiar to humans, negative effects may occur on the education system (Westland 2020). At the same time, when the scope of education turns into a lifelong education concept to include everyone, the problem of over-dependence on artificial intelligence can be encountered. In this process, human judgment, supervision, and control must be functional to detect and prevent potential ethical issues early.

2. *Robustness of AI*: It is meaningful to perform control tests on technical robustness and security to obtain the expected efficiency of artificial intelligence applications. In particular, AI systems need to be highly secure and resistant to cyber-attacks. In this process, business continuity and emergency action plans must come into play in case of problems and have a return plan for corporate sustainability (Wodechi 2019). It is a fact that there is a lot of student information registered in the database of higher education institutions. The highly sensitive data (including emotional data) in the database regarding the learning processes of the individuals in question can be attacked by malicious people. However, there is evidence in the literature that artificial intelligence systems may make wrong suggestions. For example, it may generate harmful information for innocent individuals in course choices, career planning, and behavioral matters.
3. *Algorithmic Bias of AI*: Self-learning processes and algorithms are the basis of artificial intelligence applications. There may be a possibility that the algorithmic structure in question contains bias (Wilson et al. 2017). Therefore, there may be the possibility of unfair discrimination against some student groups in the university as a result of the information generated by algorithmic bias. A similar situation may occur in many ways regarding discriminatory actions toward gender.
4. *Privacy of students' and learners' data*: Compared to the other issues, the most important is Privacy, the area affected by artificial intelligence. It is open to discussion as a fundamental right that is particularly affected by AI systems. The control and surveillance mechanism must be functional to prevent damage to confidentiality. That is also important to ensure the quality and integrity of the data used in the HE sectors. In this context, it also requires adequate data management that includes access protocols and the capacity to properly process data in the application areas of artificial intelligence systems in higher education (Messina 2018). Special attention should be paid to the protection of any data, including personal information, behavioral habits, and biometric information of students and learners enrolled in the higher education system (Massy 2016). Otherwise, these data can be used inappropriately by different segments to exploit, manipulate, or pressure students and learners. If the protection and management of

> these data are not good, it may not be possible to obtain the expected social benefit from the higher education system (Kandeh and Alsahli 2020). Therefore, it should be ensured that access to data is carried out in a highly controlled manner and in proportion to the needs defined based on the corporate governance principles of HE institutions.

The CAE is responsible for establishing the balance between digital transformation, big data analysis, and most importantly, the necessity of ethical consideration of artificial intelligence applications, and ensuring the continuity of legal regulations. To achieve this, the CAE must present the basic mechanisms that are critical to the higher education institution's governing board and explain its importance.

It is important for CAEs to inform the higher education institution's top management of these risks and challenges and to provide advice to raise the necessary awareness regarding the importance of the ethical decision-making process. At the same time, the findings obtained as a result of the field works carried out by the internal auditors should also include value-added recommendations. In this context, CAEs can assist them and aim to strengthen their ethical dimensions. Internal auditors should determine and control both ethical and regulatory frameworks of the digital audit universe (Sambamurthy and Zmud 2017). It should be noted that there is a need to tightly control how AI can be designed, developed, and deployed for higher educational purposes. The list of possible mechanisms is provided below.

Sometimes, there may be a need for additional Regulation to change, adapt, or add a new item, so that the processes regarding designing, developing, or deploying AI in Education would be forced to behave and make decisions ethically. It is important to develop "ethical codes of conduct" and to be internalized as a corporate culture in higher education institutions. The point to be noted here is that it is not enough for employees to make ethical decisions (Albrecht 2008; ACCA 2016). However, a holistic perspective is needed, including the involvement of all stakeholders in the process, including suppliers, researchers, and social groups that provide application development support for information technologies. Certification and accreditation are especially important to establish and maintain the standard of these processes.

NOTES

1 The Going Digital Toolkit helps countries assess their state of digital development and formulate policy strategies and approaches in response. Data exploration and visualization are key features of the Toolkit. For further details, visit OECD: https://goingdigital.oecd.org/en/

2 The point to be noted here is that the structure of universities and their sectoral business model is quite different from other sectors. Therefore, it is important to

review the internal audit structure accordingly. That is a challenging issue, and it is aimed to contribute to meeting a significant need in the HE sectors through this book.

3 The ethical concerns can be defined in relation to the principles of equity, clarity, transparency, privacy, and conformity.

REFERENCES

ACCA. (2016). Professional accountants – The future: Drivers of change and future skills. https://www.accaglobal.com/an/en/technical-activities/technical-resources-search/2016/june/professional-accountants-the-future-report.html

Albrecht, C. C. (2008). *Fraud and forensic accounting in a digital environment*. Provo, UT: Brigham Young University. Retrieved from www.theifp.org/research-grants/IFP-Whitepaper-4.pdf

Anissa, N., Baker, T., Smith, L. (2019). *Educ-AI-tion Rebooted: Exploring the future of artificial intelligence in schools and colleges*. England and Wales: NESTA.

Blaschke, M., Cigaina, M., Riss, U. V., Shoshan, I. (2017). Shaping the digital enterprise trends and use cases in digital innovation and transformation. Part I and Chapter 6. In G. Oswald & M. Kleinemeier (Eds.), *Shaping the digital enterprise*, Cham: Springer Publications. ISBN:978-3-319-40967-2.

Brynjolfsson, E., Milgrom, M. (2013). Complementarity in organizations. In R. Gibbons & J. Roberts (Eds.), *Handbook of organizational economics*. Princeton, NJ: Princeton University Press.

Cantú-Ortiz, F. J., Fangmeyer Jr., J. (2018). University performance in the age of research analytics. Chapter 13. In F. J. Cantú-Ortiz (Ed.), *Research analytics boosting university productivity and competitiveness through scientometrics*. London: Taylor & Francis Group. ISBN:978-1-4987-6126-0.

Chartered Institute of Internal Auditors. (2020). Impact of digitization on the internal audit activity. https://www.iia.org.uk/resources/

COSO. (2019). Managing cyber risk in a digital age. https://www.coso.org/Documents/COSO-Deloitte-Managing-Cyber-Risk-in-a-Digital-Age.pdf

CUC. (2008). *Handbook for members of audit committees in higher education institutions*. Committee of University Chairmen Publications. ISBN:1-902369-21-1.

D'andrea, V.-M. (2007). Improving teaching and learning in higher education: Can learning theory add value to quality reviews? In D. F. Westerheijden, B. R. Stensaker, & M. J. O. Rosa (Eds.), *Quality assurance in higher education*, Vol. 20. Dordrecht: Springer Publications. ISBN:13-978-1-4020-6012-0 (e-book).

Dai, J. (2017). Three essays on audit technology: Audit 4.0, Blockchain, and Audit App. PhD Dissertation. The State University of New Jersey, Rutgers. Retrieved from https://rucore.libraries.rutgers.edu/rutgers-lib/55154/

Deloitte. (2012). The digital workplace: Think, share, do Transform your employee experience. https://www2.deloitte.com/content/dam/Deloitte/mx/Documents/human-capital/The_digital_workplace.pdf

Deloitte. (2020). Data driven internal audit what to do Monday morning 9AM. Deloitte Risk Analytics. https://www2.deloitte.com/ce/en/pages/deloitte-analytics/solutions/data-driven-internal-audit1.html

Esteves, M., Matias, R., Bernardino, R., Távora, V., Pereira, A. (2019). Project based learning a new approach in higher education: A case study. In M. E. Auer &

T. Tsiatsos (Eds.), *The Challenges of the Digital Transformation in Education*, vol. 917, Cham: Springer Nature Switzerland AG. ICL 2018, Advances in Intelligent Systems and Computing, pp. 525–535. DOI: 10.1007/978-3-030-11935-5_50

EU. (2019). *Ethics guidelines for trustworthy AI*. European Commission, 2019. Independent High-Level Expert Group On Artificial Intelligence Set Up By The European Commission, European Commission B-1049 Brussels. https://ec.europa.eu/digital-single-market/en/news/ethics-guidelines-trustworthy-ai

EU. (2020a). Digital Economy and Society Index-DESI (2020). Thematic chapters. https://ec.europa.eu/digital-single-market/en/news/digital-economy-and-society-index-desi-2020

EU. (2020b). Emerging technologies and the teaching profession. Ethical and pedagogical considerations based on near-future scenarios. JRC Science For Policy Report. ISBN:978-92-76-17302-1.

Flowers, M. (2019). The elevating internal audit's role: The digitally fit function 2019 State of the Internal Audit Profession Study. https://www.pwc.ru/en/riskassurance/assets/2019-state-of-the-internal-audit-profession-study-en.pdf

Institute of Internal Auditors – IIA. (2017). International standards for the professional practice of internal auditing (Standards). IPPF-Standards-2017.pdf (theiia.org)

Jarke, J., Breiter, A. (2019). Editorial: The datafication of education. *Learning, Media and Technology*, 44(1), 1–6. DOI: 10.1080/17439884.2019.1573833

Jones, S., Lefoe, G., Harvey, M., Ryland, K. (2012). Distributed leadership: A collaborative framework for academics, executives and professionals in higher education. *Journal of Higher Education Policy and Management* 34(1), 67–78.

Kahyaoglu Bozkus, S. (2019). An analysis on the implementation of new approaches and techniques in the auditing of business processes based on blockchain technologies, Chapter 6. In B. Darici & F. Ayhan (Eds.), *Cryptocurrencies in all aspects*. Berlin: Peter Lang Publications. ISBN:978-3-631-78387-0.

Kahyaoglu Bozkus, S., Aksoy, T. (2013). Measuring the internal audit performance: Tips for a successful implementation in Turkey. *American International Journal of Contemporary Research*, 3. SSN2162-139X (Print).

Kahyaoglu Bozkus, S., Caliyurt, K. (2018). Cyber security assurance process from the internal audit perspective. *Managerial Auditing Journal*, 33(4), 360–376.

Kahyaoglu Bozkus, S., Balkan, B., Balkan, O. (2019). Determinants of ethics auditing: Structural equation model approach. In K. Çalıyurt (Ed.), *Ethics and sustainability in accounting and finance, volume I. Accounting, finance, sustainability, governance & fraud: Theory and application*. Singapore: Springer. http://doi-org-443.webvpn.fjmu.edu.cn/10.1007/978-981-13-3203-6_6

Kandeh, H., Alsahli, M. (2020). Effect of big data analytics on audit an exploratory qualitative study of data analytics on auditors' skills and competence, perception of professional judgment, audit efficiency and audit quality. UMEA University, Department of Business Administration Master's Program in Accounting Master's Thesis in Business Administration III, 30 Credits, Spring 2020 Supervisor: Tobias Svanström.

Kovalenko, D., Briukhanova, N., Kupriyanov, O., Kalinichenko, T. (2019). Academic determination of technical information optimization due to information and communication technologies. In M. E. Auer & T. Tsiatsos (Eds.), *The Challenges of the Digital Transformation in Education*, vol. 917. Cham: Springer Nature Switzerland AG 2019. ICL 2018, AISC 917, pp. 25–34. DOI: 10.1007/978-3-030-11935-5_3.2

KPMG. (2013). Taking the pulse. A survey of internal audit in Singapore 2013. https://assets.kpmg/content/dam/kpmg/sg/pdf/2016/09/sg-Audit-Internal-Audit-Survey-2013.pdf

KPMG. (2020). 20 key risks to consider by Internal Audit before 2020 are you aware of the risks concerning Internal Audit today and in the near future? https://assets.kpmg/content/dam/kpmg/ch/pdf/key-risks-internal-audit-2018.pdf

Luckin, R., Holmes, W., Griffiths, M., & Forcier, L. B. (2016). Intelligence unleashed. In *An argument for AI in education*. London: Pearson. ISBN:9780992424886.

Marr, B. (2018). How much data do we create every day? The mind-blowing stats everyone should read. *Forbes*, 21 May. https://www.forbes.com/sites/bernardmarr/2018/05/21/how-much-data-do-we-create-every-day-the-mind-blowing-stats-everyone-should-read/%20-%203eac99fb60ba/?sh=35dced8673c8

Massy, W. F. (2016). *Reengineering the university: How to be mission centered, market smart, and margin conscious*. Baltimore, MD: Johns Hopkins University Press. ISBN:978-1-4214-2274-9.

McDaniel, M. A., Anderson, J. L., Derbish, M. H., Morrisette, N. (2007). Testing the testing effect in the classroom. *European Journal of Cognitive Psychology*, 19, 494–513.

McGhee, M., Grant, S. (2019). Audit and technology. ACCA and Chartered Accountants Australia and New Zealand (CA ANZ) Joint Report, Audit and Technology | ACCA Global.

Messina, M. (2018). Designing the new digital innovation environment. In G. Bongiorno, D. Rizzo, G. Vaia (Eds.), *CIOs and the digital transformation*. Cham: Springer. DOI: 10.1007/978-3-319-31026-8_9

Milgrom, P., Roberts, J. (1995). Complementarities and fit strategy, structure, and organizational change in manufacturing. *Journal of Accounting and Economics*, 19(2–3), 179–208.

OECD. (2019). *Going digital: shaping policies, improving lives*. Paris: OECD Publishing. DOI: 10.1787/9789264312012-en

OECD. (2020a). Education and skills today. https://oecdedutoday.com/

OECD (2020b). *Education at a Glance 2020 – OECD Indicators*. Education at a Glance 2020 (windows.net)

Porter, M. E. (1996). What is strategy?. *Harvard Business Review* 74 (60), 61–78.

PwC. (2012). Aligning internal audit. Are you on the right floor? State of the internal audit profession study. https://www.pwc.com.tr/en/risk-surec-teknoloji-hizmetleri/assets/ic-denetim-ve-kontrol-hizmetleri/kuresel-ic-denetim-anketi-2012-en.pdf

PwC. (2019). Elevating internal audit's role: The digitally fit function 2019 State of the internal audit profession study. https://www.pwc.ru/en/riskassurance/assets/2019-state-of-the-internal-audit-profession-study-en.pdf

Ramamoorti, S., Siegfried, A. N. (2016). Promoting and supporting effective organizational governance. Internal Audit's Role. The Institute of Internal Auditors Research Foundation (IIARF). ID #2016-0423.

Risk Management Society (RIMS) and Institute of Internal Auditors (IIA) (2012). Executive report | Risk management and internal audit: Forging a collaborative alliance. https://global.theiia.org/standards-guidance/Public%20Documents/RIMS%20and%20The%20IIA%20Executive%20Report%20Forging%20a%20Collaborative%20Alliance.pdf

Sambamurthy, V., Zmud, R. W. (2017). *Guiding the digital transformation of organizations*, 2nd ed. Legerity Digital Press LLC. ISBN:978-0-9995347-0-0.

Siggelkow, N., Rivkin, J. (2005). Speed and search: Designing organizations for turbulence and complexity. *Organization Science*, 16 (2), 101–122.

Stieglitz, N., Foss, N. J. (2015) Business model innovation the role of leadership, Chapter 6. In N. J. Foss & T. Saebi (Eds.), *Business model innovation the organizational dimension*. Oxford, UK: Oxford University Press. ISBN:978-0-19-870187-3.

Tushman, M. L., Anderson, P. (1986). Technological discontinuities and organizational environments. *Administrative Science Quarterly*, 31(3), 439–465.

UNESCO. (2019). Artificial intelligence in education: Challenges and opportunities for sustainable development (2019). ED-2019/WS/8. https://unesdoc.unesco.org/ark:/48223/pf0000366994

Vuorikari, R., Punie, Y., Cabrera, M. (2020). Emerging technologies and the teaching profession: Ethical and pedagogical considerations based on near-future scenarios, EUR 30129 EN, Publications Office of the European Union, Luxembourg, ISBN:978-92-76-17302-1, DOI: 10.2760/46933, JRC120183.

Westland, J. C. (2020). *Audit analytics. data science for the accounting profession*. Switzerland: Springer Publications. ISBN:978-3-030-49091-1.

Wilson, H. J., Daugherty, P. R., Morini-Bianzino, N. (2017). The jobs that artificial intelligence will create. *MIT Sloan Management Review*, 58(4) Online.

Wodechi, A. (2019). *Artificial intelligence in value creation improving competitive advantage*. Switzerland: This Palgrave Macmillan. ISBN:978-3-319-91595-1.

Chapter 4

Establishment of the internal audit function balancing between physical and digital

4.1 INTERNAL AUDIT VS. DIGITAL AUDIT

When the development trend of the last decade is explored, it is observed that the audit and quality assurance activities in the higher education sectors have undergone change and transformation as in many other sectors (European Commission-Education, Audiovisual and Culture Executive Agency (EACEA) 2020). Especially with the use of advanced technologies such as artificial intelligence (AI) and machine learning, blockchain, and data analytics in internal audit, a significant transformation is emerging in the audit profession (Kahyaoglu Bozkus 2019; Salijeni et al. 2019). While internal audit used to be based on a reactive and backward-looking practice with a traditional approach, it now transforms internal audit field works by turning it into a more proactive, continuous, and forward-looking source of insight (Edwards 2012). Therefore, technology has reached such a stage that it offers internal auditors the opportunity to serve better and produce higher quality audits, even in the context of a traditional internal audit perspective (Abraham 2013).

In general, risk is common in every business environment. When we consider any transaction and business process that a university undertakes, for each of them, it requires identifying and managing the risks that could hinder the execution of the transactions and processes in an effective, economical, and efficient manner. It is also important for a university to be alert not only to internal operational risks, but also to external risks arising from advances in technology, changing economy and market conditions, and increasing globalization and digitalization trends in HE sectors (University Risk Management and Insurance Association (URMIA) 2007).

Higher education institutions are facing these challenges more and more every day, and it appears that they are under increasing pressure from the government, public, and campus community to manage existing and emerging types of risk (NACUBO and AGB 2007; Association of Governing Boards of Universities and Colleges (AGB) and United Educator (UE) 2009). Within the framework of the corporate functioning and business model of HEIs, it should manage risks in the physical and virtual environment in a

DOI: 10.1201/9781003093008-4

wide variety of areas such as safety and security, regulatory compliance, academic affairs, research, information technology, finance, human resources, and facility management (Abraham 2013). In addition, recent events such as COVID-19 that caused the global economic crisis and social problems such as malfunctions in the campus, student actions indicate how important and priority risk management is in higher education. In reality, although the institution may survive such events, the same may not be possible for the leadership structure of the university (Tufano 2011; McCaffery 2019). In this context, it is critical to have a functional internal audit department evolving with highly digital talents.

A study by Brewer and Walker (2011) reveals new risks that universities are exposed to. These authors point to the new era by explaining that the higher education context in which universities operate consists of a complex network of education providers in a competitive global virtual environment. Accordingly, in the new higher education sector environment, several important change factors affect a university's activities and lead to risks (Ariff et al. 2014). These change factors should be closely monitored by internal audit departments and their possible effects should be examined and added to internal audit field works. Thus, the main factors that internal auditors should consider can be classified as follows:

Increasing expectations in the field of responsiveness and accountability: Higher education institutions generally need to take into account the needs of employers and the wider society in the design and delivery of courses that constitute their main field of activity. In addition, there are increasing expectations for responsiveness and accountability regarding the research studies conducted through university research centers (Soares et al. 2016). This situation is becoming more and more important to audit such activities and processes systematically and periodically.

Expansion of the scope of student expectations: Considering the current learning paradigms, it is a fact that the corporate brand and image gain importance due to the demonstration of the quality of education and research offered by a university (Kageyama 2014). That also concerns the reputation of the university. In this way, it enables high-quality students and faculty to attract and respond to expectations. Internal auditors should consider and determine the risk factors affecting the corporate image and reputation of the HE institution.

Increasing competition: The higher education sector is now becoming a global market with the effect of digitalization. There is increased competition for students and faculty in such a huge education market. The source of competition in faculties is becoming more evident due to the demographic shift toward higher age groups as target learners and consequently difficulties in sequential planning. This situation may lead to operational risks which internal auditors are forced to monitor and determine via early warning mechanisms via CA/CM techniques (Zhang et al. 2015).

Increased external scrutiny: Activities carried out in higher education institutions are now subject to increasing scrutiny and evaluation compared to previous periods (Yilmaz 2013). This makes it necessary for internal audits to closely monitor regulatory and stakeholder requirements and ensure compliance.

Entrepreneurship: It is a fact that the interaction of higher education institutions with the commercial world has increased. This practice is an integral part and increasing feature of most universities' activities. To be an entrepreneurial and innovative university, active involvement entails proactively seeking partnerships with business organizations where research work can be financed and the commercial value of research output realized and shared (Abrash 2017). That is guided by university researchers who demand regulations that support entrepreneurship and an innovative approach to the business world. Therefore, it is necessary to control and periodically monitor the risk level of such works and interactions, which arise from the outward-looking structure of higher education institutions and research centers and are carried out with cooperation protocols (Abbott et al. 2016).

Impact of information technology: Advances in information technology have affected higher education in two main areas: first, the widespread emergence of digitalization or e-learning platforms. In this context, remarkable changes occur in traditional learning and teaching models. While the increased flexibility in the new training model offers an advantage, it results in a new risky area added to the risk universe in terms of audit (Salijeni et al. 2019). The point to be considered here is to analyze the possibilities and background risks exposed by direct communication with the learners and to manage the risks by early diagnosis. Secondly, it is possible to automate student support services such as student admissions and registration through infrastructure improvements provided by technological developments. In addition, a common digital communication platform can be created that enables the integration of financial and other administrative functions (Abbott et al. 2016).

In the study conducted by Ruzic-Dimitrijevic and Dakic (2014), the reasons for the emergence of "academic risks" in British universities were revealed. The authors explain sources of academic risks as follows:

Academic excellence: To ensure this, universities need to employ adequate staff and students. However, disruption in employment for research in universities causes insufficient infrastructure or poor rankings. Internal audit is responsible for monitoring the complete and timely realization of the objectives to be met so that academic excellence does not remain unfulfilled and is fully implemented. In this context, it should be ensured that the necessary control points are created for the goals determined by the university senior management to be realistic and to be completed within the specified periods. The internal audit department can act as a "trusted advisor" in this process and offer value-added recommendations contributing to the performance increase in the HE institution. For this, it is necessary for the internal audit team to closely monitor technological developments, analyze the sectoral business risks well, and give importance to professional development and training with the utmost care.

The overall quality of the higher education sector: As factors that negatively affect the level of quality, poor leadership, insufficient uncertain goals, or insufficient evidence of the performance of the higher education institution can be cited.

Performance of higher education: Insufficient demand structure for universities, insufficient representation of socio-economically diverse communities, or unsuitable demands for national educational capacity may lead to a decline in performance indicators of HE sectors. It is a fact that there are issues that distinguish the higher education sectors from other sectors and are imperative for auditors. These issues can be defined as "academic risks". For the performance of universities to reach the targeted level, it should be taken into consideration that HE sectors have different dimensions compared to other sectors. In this respect, Wang (2010) argues that it is practical to distinguish between the academic function of the university and the performance related to the management function. Within the framework of this approach, two focal dimensions should be taken into account when examining and measuring the performance of universities. Academic activities and management activities are closely interlinked in university practice. Both fields of activity are key factors in the performance of universities. The point to note here is that a university may have an excellent management system with distinguished staff, but this does not necessarily mean that exquisite performance can be achieved in the university as a consequence (Barnabe and Riccaboni 2007; Adriana et al. 2008).

In this respect, "academic performance" is the core of a university's performance as a top priority area overall. Management performance, which is in a supportive position within the university, can be expressed as a performance that can increase academic performance and serve it (Broadbent 2007). Together, they help create a complete picture of performance for the university concerned while fulfilling different functional roles at the university. When evaluated in this respect, it may be necessary to clearly distinguish between academic and "management performance" in measuring performance in universities. This distinction also contributes to the development of a measurement approach specific to university environments for internal auditors.

In the traditional risk management approach, risks are generally monitored and managed one by one. However, since the diversity of risks that higher education institutions are exposed to as described here increases, it is impossible to successfully manage risk with such a traditional approach. Therefore, ERM techniques that present an integrated perspective are needed (Clyde-Smith 2014; Lundquist 2015; COSO 2019).

The necessity of more advanced technologies with integrated risk management is obvious. In this context, the traditional physical fieldwork of internal audit is replaced by remote monitoring and auditing of big data in the virtual environment. In this new audit approach, which we can describe as "digital audit", a method that differs much from the conventional internal audit is required. Especially with the changing nature of the audit universe, i.e., the "digital audit universe", it is crucial for internal audit to transform the business models and business processes accordingly. It should be noted that it is impossible to conduct an audit of a university's education and research environment with digital ecosystems using traditional internal audit tools and techniques and to give assurance to the HE senior management.

Another feature required by the digital audit is that internal auditors have to choose and use appropriate and efficient approaches and tools that are convenient for higher education sectors business model to solve a wide variety of problems accurately, i.e., "problem-solving skills". As fundamental indicators of the overall problem-solving capabilities of internal auditors, skills such as "Information Gathering", "Problem Recognition", "Creative Thinking", "Systems Analysis", and "Decision Making" are required to cover all processes from planning to the last stage of the audit.

In summary, these practices cause the transformation from internal audit to digital audit to be a fundamental requirement in the higher education sector. In the transition from internal audit to digital audit, care should be taken to be versatile while demonstrating the competencies of internal auditors. Especially the need for success stories about communication and collaboration will become more important in the future. In this context, since virtual environments replace physical environments, which are important in communication, internal auditors will have the responsibility to convey to the other party in the most effective way, the thoughts, ideas, data obtained from fieldwork, information, and messages.

At the same time, the internal audit charter, which is the official communication tool of internal audit, should be harmonized with this digitalization process requirements and announced to all relevant stakeholders. Ramamoorti and Siegfried (2016) state that internal auditors have "twin goals" related to the corporate governance process of an organization. These are as follows: First, "promoting the value preservation" and second, "supporting value creation efforts". These twin goals of internal auditors are directly related to the governance structure of the organization leading to strategic growth and success criteria often preferred by executive management. In this respect, the "internal audit charter" should be in line with these twin goals to position the internal audit department as having a seat at the table.

4.2 THE INTERNAL AUDIT CHARTER: REALITY VS. FICTION

Considering the digital audit universe in such rapidly changing and developing HE sectors, internal auditors need to decide most assuredly where to start and determine their strategic directions. The place where this strategic direction needs to be laid out and detailed is the internal audit charter. The point to note here is to determine whether the education is fully transferred to the online platform and to analyze whether the university education system includes simple learning techniques or more complex structures. Thus, it may be possible to determine the level of corporate maturity and form and implement a risk-based audit plan accordingly.

An internal audit charter should include clear objectives and responsibilities expected from an internal audit (Pitt 2014). From this point of view, the internal audit charter should comprise the internal audit's business strategy as a complementary element that supports the corporate strategy of the HE institution. Haynes et al. (2004) state that there are various models and levels of online learning in the HE sector. It should be taken into account that hybrid systems (also known as blended learning or augmented learning) can be found in addition to full online systems in this regard, which is necessary for internal audit professional practices. That is imperative when expressing the essential fields of responsibilities of internal audit in the charter. Perhaps most importantly, an internal audit charter is a prerequisite for an internal audit to be formally positioned within the HE institution as an indicator of its unbiasedness, independence, and objectivity.

It is a fact that under the umbrella of hybrid learning practices in HE institutions, there will be remarkable diversity in the mix of face-to-face learning methods and online methods that are used. We discuss the effect of this mix concerning auditing techniques further throughout the book. In this context, the essence of what is happening is that elements of the new online approach to distance learning are converging to augment traditional face-to-face learning styles. This offers new opportunities and challenges to stakeholders in the HE sectors in order to encourage self-directed learning mechanisms all over the world. Hence, an internal audit is not an exception to this learning process.

To effectively oversee and audit the online learning processes in the universities, internal auditors are required to have some self-disciplinary and initiative qualifications. In this respect, they need to develop a work program and stick to it to execute their professional work with high performance.

It is a fact that all internal auditors may not have developed these qualifications and digital skills well when they first start their role in the university, and they may even opt for traditional face-to-face audit field works if they have to choose (Holt 2012). From this point of view, it is necessary to include basic needs and expectations in line with the institutional structure within the scope of audit guidelines, policies, and procedures. The main determinant of this and the open communication tool with stakeholders in the internal audit charter.

A study by Burch (2011) emphasizes the importance of taking into account the expectations of stakeholders, especially in the organization and the transformation of the internal audit department. It is important that these expectations are understood correctly and at the same time reflected in the internal audit charter, which is the official communication tool. In this context, it should confirm that a chief audit executive should understand how the organization intends to engage with internal audit and how it wishes to benefit from it in the initial phase and in interim periods that require revision. This should help define the internal audit's role that needs to be approved by senior management and the audit committee so that the verification process is transparently communicated to all stakeholders beforehand (Aksoy and Kahyaoglu 2012).

According to Pitt (2014), it is important to take the strategic and organizational structure as a basis for the positioning of the internal audit department within the HE institution. First, the scope of organizational structure is explained as follows.

The true understanding and analysis of the environment in which higher education institutions are located and in which they operate is a fundamental requirement for internal auditors to add value. It should be ensured that this understanding and analysis are objectively reflected in the internal audit charter. In this context, the main indicators of the HE organizational environment include:

1. Higher education sectors regulatory and policy environment
2. Political environment
3. Key drivers and strategic goals of HE institution
4. Key competitors
5. Potential opportunities and challenges in the HE markets
6. Key customer and customer demographics

These contextual indicators determine the types of results and outputs expected from the internal audit function, as well as the activities that the internal audit should undertake.

Second, the scope of strategic structure is explained as follows: The strategic function of internal audit is influenced by both the broad organizational context and the strategic context of the internal audit itself. As an organizational context, the legal regulations to which the HE sector is subject on a strategic basis, the global competitive environment, and education policies can be given as examples. However, the strategic context of internal audit defines the motivation behind internal audit's existence (Holt 2012). Accordingly, if there are legal requirements for the internal audit establishment, this must be clearly stated in the internal audit charter and reference should be made to the underlying legal regulation. In addition, the internal audit charter should include the main challenges faced by the internal audit function and what value it aims to provide to the stakeholders with a strategic perspective (IIA 2019a).

The internal audit charter should describe the structure of assurance services provided to the organization. At the same time, consulting activities are specifically defined. Clear disclosure of this information is important as it is used to measure the quality of internal audits (Galloway 2010; Aksoy and Kahyaoglu 2012; Pitt 2014).

The internal audit should take into account not only internal obligations but also external obligations when performing its duties in a highly regulated higher education market. In this respect, the internal audit has an important role in full compliance with the legal legislation that the higher education sector must comply with. Specifically, it would be appropriate to have clear provisions in the internal audit charter to detail the necessary audit types, and business processes.

For the internal audit department to perform the necessary value-added approach within the higher education institution, it is decisive how the communication and reporting infrastructure with the Audit Committee (AC) will be (Pickett Spencer 2012). In this context, it should be ensured that detailed information is included in the internal audit charter.

According to the study by Rezaee (1996, 2010), the quality of communication with the AC is necessary to strengthen the independence and impartiality of internal auditors. Therefore, it argues that the internal audit function can be strengthened if internal auditors have the opportunity to communicate their findings and opinions directly to the AC.

In the position paper prepared by the IIA (2019b), seven main topics that determine the effectiveness elements related to internal audit and which are recommended to be included in the internal audit charter are determined. However, considering the structure of digitalized higher education institutions, it is meaningful to examine the scope and relevance of the internal audit charter proposed by The IIA and to implement the necessary revisions within the framework of technological requirements. In this context, discussion regarding the internal audit charter framework according to IIA standards and the revision areas arising from the digital audit are presented below. The key issues and challenges are summarized as follows:

1. *Mission and Purpose of Internal Audit* – First of all, the internal audit charter should define both the mission and the purpose of the internal audit function. Overall, the mission should provide a strategic

perspective to increase and protect the brand value of the relevant higher education institution by providing risk-based and objective assurance, advice, and insight. However, it is recommended that internal audits be designed to add value and improve the organization's operations through independent and objective assurance and consulting services.

In defining the mission, vision, and purpose of internal audit, it is necessary to convert their perspective and enter more advanced areas in this HE sector. Therefore, it is expected that internal audit will continue to renew and continuously monitor the Internal Audit Charter to ensure that its efforts to maximize value are realistic. It should be noted that the perception of the internal audit function should no longer be a "Cost Center" and rather targets should be set to become a "Profit Center" instead.

In the internal audit charter, the effect of technological developments and innovations on the value generation process should be defined. This definition should be internalized by the internal audit team to apply it with a systematic and disciplined approach based on IIA standards. An example of this systematic approach for defining the mission and purpose of internal audit is shown in Figure 4.1. It should be noted that internal audit is not only expected to maintain the existing corporate value but also add to the organizational value with an integrated approach. Therefore, excessive focus on only one of these areas can lead to deterioration and unbalance of audit value. To prevent this situation, value management should be the priority of CAEs. In this way, the internal audit charter will provide the basis for the "reality" of audit fieldworks rather than the "fiction".

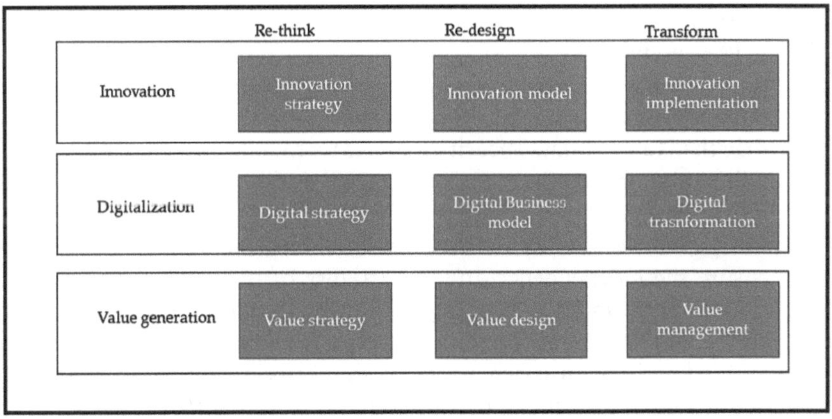

Figure 4.1 Re-thinking internal audit mission and purpose in HE sectors.

Source: Adapted from Schreckling and Steiger (2017).

2. *International Internal Audit Professional Practice Standards* – The internal audit charter should provide detailed information on how the internal audit function is managed in the general framework. It would be appropriate to include statements stating that the IIA adheres to the "International Professional Practices Framework (IPPF)".
   - Standards
   - Basic principles for the professional practice of internal audit
   - Definition of internal audit
   - Ethical rules

   Increasing the "Internal Audit Digital Maturity Model"[1] and reaching the capacity to audit by internalizing digital developments can be expressed as an important challenge.
3. *Authorization and Reporting Lines* – The CAE's operational and administrative reporting lines and relationships in the organization should be described in detail. In addition, a statement confirming that the governing body has sufficient authority throughout the organization to perform the duties of the internal audit function should be included.

   It should be noted that according to IIA standards, written reporting is not required to present internal audit findings. Standard 2440 states that within the scope of "Dissemination of Results", the chief audit executive must convey the results to appropriate parties in the HE institution. In the practice guide, it is stated that internal audit results can be communicated orally or in writing. This communication format can be changed according to the recipient. In this context, PowerPoint and Prezzi presentation software tools commonly used by internal auditors have been relevant in the market for years. However, new data-connected tools like SharpCloud mean that digital business leaders may request more tech-savvy reporting in the HE sectors.

   Presentation of the findings obtained as a result of the digital audit to stakeholders related to digital reporting becomes a part of the internal audit business model. In connection with this, it should be noted that the issues of data security and access to data and data ownership will gain more importance. The internal audit charter should include information regarding these issues and express responsibilities.
4. *Independence and Impartiality* – It should be stated that the CAE will ensure the independence and impartiality of the internal audit function to perform its duties autonomously. In addition, internal audit should not have direct operational responsibility or authority over any audited activity. In other words, it should be stated that an internal audit is not an executive unit but only a "trusted advisor".
5. *Scope of Internal Audit Activities* – Define the scope of the internal audit function in line with the organization's strategic goals and

objectives. The scope should include providing independent assessments of the adequacy and effectiveness of governance, risk management, and control processes.

The scope of internal audit is shifting from the physical environment to more and more virtual environments compared to the past. It is implicit that this new environmental change is addressed and explained in the internal audit charter. The main point here is that the digitization level of each department and process may not be the same considering the scale of organizational structure. Therefore, it may be indispensable to make a preliminary assessment of whether it is suitable for digital audit or CA/CM techniques. It is necessary to closely monitor the units and processes that can be audited digitally and those that cannot be done. Accordingly, while the digital audit is possible in departments and processes that complete digital transformation, other departments and processes may need to benefit from internal audit consultancy services. It should be analyzed which services the audit will provide according to the maturity level within the organization, that is, whether assurance or consultancy can be provided.

6. *Responsibility* – The overall responsibility of the internal audit function should also be defined in the internal audit charter. An internal audit charter prepared under IIA standards is required to perform the following at least once a year:
   - A risk-based internal audit plan
   - Verifying that the internal audit activity has access to appropriate, competent, and qualified resources
   - Verifying that the internal audit function is performing its task
   - Assurance of IIA standards
   - Communicating the results of its work and tracking agreed corrective actions
7. *Quality Assurance and Improvement Program* – Define internal audit's Quality Assurance and Improvement Program (QAIP), which covers all aspects of the internal audit function and under IIA standards:
   - Compliance with IIA Standards and reporting QAIP results to senior management periodically
   - External evaluation is required at least every five years

Since generating value based on change, technology and innovation come to the fore in the new audit approach, the quality assurance review (QAR) process becomes a critical issue. Within the framework of IIA standards, it is recommended that this process be implemented based on external evaluation within five years at the latest. Considering the speed of technology, it would be appropriate to apply it more frequently than to the last moment. It is also important to administer more internal evaluation and self-assessment techniques without waiting for external evaluation. Thus, an audit recommendation and contribution based on a cost–benefit analysis can be presented.

4.3 REALITIES OF EMBRACING RISK-BASED AUDITING IN DIGITAL SPACE

The processes and controls of a higher education institution at a definite moment in time can be efficient, economical, and effective for today. However, this situation may not be the same in the future, and when risks are not identified in time, their processes can be jeopardized. For example, the use of e-learning technology instead of conducting classroom training previously carried out by academics brings new risks. Therefore, higher education institutions should adjust their processes and controls accordingly to avoid disadvantages and challenges. In this context, IA can contribute to understanding the impact of automation and changing team skills and dynamics, and in this case, internal audit can play an important role in risk management and decision-making processes in the HE sectors.

It can be stated that internal auditors need to be more "flexible" than ever to help analyze the risks faced by the higher education institution and their impact on the organization. From this point of view, creating an internal audit plan based on a dynamic risk assessment is no longer an option for internal an audit. Especially if IA wants to have a seat at the table in such a competitive business environment, these become a must.

However, given the digitalization process of higher education institutions and the sector in general, more and more are demanded from internal auditors. The senior management is asking IA to take a more advisory role in proactively looking at the design of controls in areas such as systems development, new product and service development, and strategic operations of the university. And, considering their broad mandate, IA is in a unique position to generate value-added works.

It is a fact that the next step for internal auditors is to apply AI and ML algorithms to improve the quality of data analysis and risk assessment and increase the rate of fraud detection in the HE sectors. Therefore, the effective use of new technological applications, tools, and techniques is needed based on risk-based auditing in digital space and structure. Within the framework of these new technological developments, the points of change in audit tools and techniques are summarized below.

The audit universe, which is the subject of internal audit fieldwork, is changing rapidly. Therefore, the scope, frequency, and flexibility of the audit are transformed within the framework of big data analysis, artificial intelligence, robotics, and blockchain applications.

In a study conducted by Vasarhelyi et al. (2020), because the tools and techniques used by internal auditors are becoming smart systems, these authors have started to define internal auditing as "smart auditing". Another point emphasized by these authors is that these audit tools and techniques do not only make internal auditors smarter. It also enables internal auditors to focus on tasks that increase their productivity by using time effectively and analyzing data more comprehensively.

On the other hand, there are also challenges posed by digital audits. That is, internal auditors now have more knowledge within the steadily growing HE institutions. Accordingly, under the presence of increased information and big data, more information should be verified in a shorter time for decision-making and assuring senior management. The most important requirement of digital audit for data validation in a short time is to standardize the data (Moffitt et al. 2018; Cohen et al. 2019). Without this standardization, there may be some practical difficulties arising in the use of audit tools and techniques.

Besides the standardization of the data, another important challenge arises from the gray areas of algorithmic forecasting methods of the data. In other words, within the framework of artificial intelligence applications, some machine learning methods can make predictions of future events based on past data. ML methods can determine patterns and extract features from big data and at this point, there is a possibility of producing a "biased result" like a "black box" nature (D'Onza and De Santis 2020).

4.4 CHALLENGES OF MEETING THE IIA STANDARDS WHEN WRITING REPORTS ON DIGITAL RISKS

For internal audit, it is difficult to explain to auditees and business unit managers how an algorithm reaches its decision based on ML and big data analytics. This makes it less appropriate to be regarded as a piece of audit evidence by today's audit standards, especially when the objectivity of the internal audit remains highly dependent on a machine-generated result. In order for smart audit techniques to be accepted and become more effective, it is necessary to standardize the big data subject to audit (Tang and Karim 2017). In addition, it is important to clearly define and implement the authorities and responsibilities regarding data management in the HE institution.

In general, it can be stated that machines are inherently better than internal auditors at performing auditing tasks defined as repetitive and rules-based. In this context, when internal auditors do not need to spend most of their time on performing frequently repetitive and routine tasks, they can focus their efforts on more challenging and critical tasks, particularly those involved in assessing the risk of material misstatements. This approach indicates the future expectations of work based on "human–machine cooperation". Zhang (2019) reveals that this can be the future work form of auditing as an important area of improvement that can be achieved.

It is a fact that the use of data analytics is not really a new topic and has been around for decades. However, based on the evolution of digitization, the increasing computational capabilities and the emergence of external data completely change the application framework of assurance services based on real-time data analysis rather than passive and post-acquired information. This new situation not only benefits the internal audit processes (Brown-Liburd and Vasarhelyi 2015). It also contributes to the development of more effective digital workflows and business culture, as it significantly increases the accuracy of transactions.

There is an important difference that separates data analytics, which is included in the internal audit of HE institutions, from other sectors. This is because not only the analysis of internal data, but also the analysis of external data generated from all learners served by the university coming through social networks has the potential to provide important information and evidence. According to the research of Appelbaum et al. (2017), analysis of such external data supports reducing the "audit risks" (Bozkus Kahyaoglu 2020) and making the findings more effective and reliable (Yoon et al. 2015).

According to Rozario and Thomas (2019), digitalization can sometimes be seen as a major concern for internal auditors. In particular, evaluations are made on whether these disruptive technologies will completely replace auditors. In this regard, technology cannot yet fully replace human auditors, but it has been shown to perform specific and narrowed tasks more effectively than human auditors. These practices are expected to continue in the future.

Digitalization can be expressed as the most important reality of today and the situation that should be accepted. In this process, difficulties arise in the acquisition, verification, and analysis of big data required by digital audit for internal audit to report in accordance with IIA standards. The first step of digital audit is the automation of data collection and data preparation. This may contain automation of fieldwork based on the documentation of working papers and processes by establishing a single process for all information exchanges between auditor and auditee.

Considering the blockchain technologies, this process may even start with the evidence, distributed ledger technologies (blockchain). It should be noted that the moment they are generated, they can be used as audit evidence. This would provide a valuable opportunity to introduce "control by design" in the strategic management process of HE institutions to establish CA/CM systems (Brown-Liburd and Vasarhelyi 2015; Bozkus Kahyaoglu et al. 2020).

On the other hand, it is also possible to automate audit fieldwork activities for the data preparation stage. That may include verification, extraction, and processing of data related to the risk-based audit plan. In this case, an internal auditor can be assisted by software, which is called "robotic process automation or RPA", that can be programmed to accomplish basic routine, and repetitive audit works in applications and data sources that can mimic human auditors' actions (Kokina and Davenport 2017).

When considering an internal audit process in general, the most important for an internal auditor under IIA standards, with or without digital audit, is to rely on the quality of the input subject to the audit. In particular, what is expected of an auditor is to take care that all the evidence she collects is relevant, reliable, complete, and up-to-date. Given the structure of the higher education sector, data entries are generally digital. The data in question is in digital media and typically held in IT systems (databases or others). In addition, "Rules" can generally be implemented through code. Business processes and decisions can be made by machines based on algorithms (KPMG 2020). Therefore, this emerging digital audit universe requires the inclusion of new roles in the audit process.

It is a fact that data and data quality is the key to successful audit fieldwork. Therefore, in the first phase of internal audit fieldwork, "data auditors" are needed to examine these data, reconcile them with materiality, and verify that they are fit for the purpose, the analysis they are intended to use (Bialik 2019). On the other hand, as the second step, the internal auditor should determine that algorithms must be transparent and explainable. Therefore, there is a need for "algorithm auditors" who can verify that the process and decision rules are properly applied and produce predictable results (Deloitte 2019).

The terms of reference of the "algorithm auditor" can be briefly expressed as follows. Firstly, it is related to exploring what factors are associated with the results of the AI algorithm and to applying the necessary tools and testing techniques to determine whether there is an inherent bias in the underlying data. However, the most important area of responsibility for the algorithm auditor is that she continues to conduct periodic reviews to determine the fairness of a model after deployment (Sandvig et al. 2014). These reviews include monitoring and deterrence activities to control black-box issues, algorithmic bias, privacy protections, and illegal discrimination (Eslami et al. 2017). Another task expected from algorithm auditors is to define the problems related to the algorithm they examine, to provide suggestions for making the model more ethical and accountable. An important area of added value expected from algorithm auditors is to help regulate how cognitive technologies make decisions and help protect society from the potential negative effects of technology. In particular, they periodically review the algorithms to ensure confidentiality and freedom from bias and discrimination (Deloitte 2019).

Subsequently, internal auditors should control that these data and algorithms are typically stored in a secure IT system. Therefore, there is a need for more "IT auditors" who can go deeper and evaluate systems, processes, and outputs by examining the data with advanced analysis techniques.

It is not an easy task to create an internal audit team with the IT competencies to talk about a faster and more efficient digital audit. It is important to specify these job descriptions in detail in the internal audit regulation and to recruit accordingly, and keep the technical information of internal auditors up-to-date with continuous training activities. All these issues can be expressed as the main challenges that internal audit is expected to solve in the future in the HE sector to sustain a standard of work in line with IIA standards.

According to De Santis (2016), improving audit quality should be the main goal of auditing standards. However, it is a known fact that the internal audit profession is full of standards expected to be followed at every stage (Knechel 2013). The point is that internal auditors should not confuse tools and objectives. In other words, standards should only represent a path to a goal. That is, this should demonstrate the audit quality. Hence, internal auditors should be concerned about the quality of their work rather than the degree of compliance with standards. This is especially relevant and true today. Because in a rapidly changing audit environment, regulations are always lagging, and various audit standards do not fully reflect the new technological environment in which auditors conduct fieldworks (Chan et al. 2018).

4.5 OUTSOURCING, CO-SOURCING, AND IN-SOURCING KEYS AND CHALLENGES IN DIGITAL

Within the framework of the internal auditor job descriptions explained above, it reveals that in terms of corporate resource planning, "outsourcing", "co-sourcing" should be among the options, as well as "in-sourcing". These three major forms of cooperation with IA departments are shown in Figure 4.2. The most important reason for this is that the number of internal auditors with high digital skills is quite low in the HE sectors, and needless to say, the costs of a competent internal auditor to the HE institution are quite high (KPMG 2020).

Differences may arise between the organizational maturity level of the internal audit and the institutional maturity level of the auditee. In such a situation, outsourcing or co-sourcing may be necessary to accelerate the alignment in terms of strategic direction, taking into account the expectations of stakeholders (Lundquist 2015). When the organizational maturity level of internal audit lags, it may turn to outsourcing or co-sourcing in different models under IIA standards. What is important is that whatever model is preferred, an internal audit should demonstrate its ability to generate added value by achieving its strategic goals. In such a decision stage, the chief audit executive needs to have leadership skills in making consistent decisions (Martino et al. 2019).

According to e D'Onza and De Santis (2020), there are two important contributions of CAE leadership. The first has a key role in providing the necessary conditions for obtaining sufficient financial resources to attract internal auditors who have the digital skills and competencies to achieve the digital transformation needed in a higher education institution. Second,

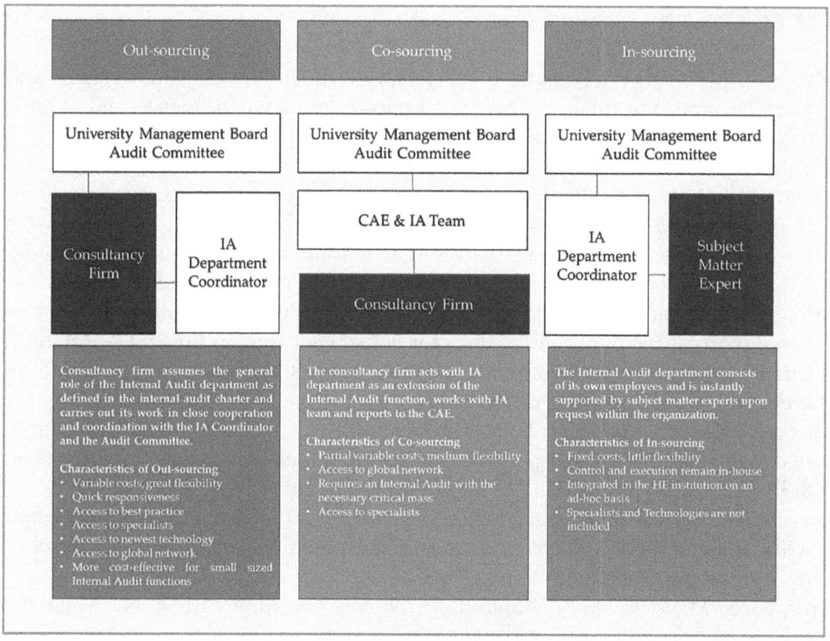

Figure 4.2 Forms of cooperation with internal audit department in HE sectors.

Source: Adapted from Pitt (2014) and KPMG (2020).

when the CAE has leadership skills, it may be easier to establish positive relationships with key actors within the HE institution. This approach is crucial for ensuring that the IA plays a central role in the digital transformation processes of the HE institution (Holt 2012).

Looking at the IIA CBOK (2015) survey data, it does not lead to an immense change in the scale of internal audit departments, even if the enterprise-scale grows too large. The average number of internal auditors was determined to be around 10 in the evaluations made on a sectoral basis. However, it can be stated that as the scale grows, the IT investment and level of development in internal audit tools and techniques increase rapidly (KPMG 2020). Because, as the scale grows, it is seen that the fieldwork structure of internal audit generally shifts to remote audit techniques, i.e., CA/CM. In cases where the internal audit team is lacking the necessary skills, priority is given to creating solutions through outsourcing or co-sourcing instead of hiring a new internal auditor (Cangemi 2015).

NOTE

1 The detailed information and a sample self-assessment model are provided in Appendixes 1–6 to measure the "digital power and skills of internal audit".

REFERENCES

Abbott, L. J., Daugherty, B., Parker, S., Peters, G. F. (2016). Internal audit quality and financial reporting quality: The joint importance of independence and competence. *Journal of Accounting Research*, 54(1), 3–40.

Abraham, J. M. (2013). *Risk management: An accountability guide for university and college boards.* Washington DC: Association of Governing Boards of Universities and Colleges and United Educators.

Abrash, L. (2017). Gaining insights from innovations in a financial statement audit. *CFO Journal* 1–6. (CFO insights & analysis written and compiled by Deloitte.)

Adriana, T., Reka, T. V., Vasile, C., Alexandra, M., (2008). Implementing the balanced scorecard in public institutions for higher education in Romania—An innovative project. *European Journal of Management*, 8(2). http://findarticles.com/p/articles/mi_6772/is_2_8/ai_n31029532/

Aksoy, T., Kahyaoglu, S. (2012). Establishment of effective internal audit function: Recommendations for best practice. *Journal of Modern Accounting and Auditing*, 8(9), 1283–1290. DOI: 10.17265/1548-6583/2012.09.003

Appelbaum, D., Kogan, A., Vasarhelyi, M., Yan, Z. (2017). Impact of business analytics and enterprise systems on managerial accounting. *International Journal of Accounting Information Systems*, 25, 29–44.

Ariff, M. S. M., Zakuan, N., Tajudin, M. N. M., Ahmad, A., Ishak, N., Ismail, K. (2014). A framework for risk management practices and organizational performance in higher education. *Review of Integrative Business and Economic Research, Society of Interdisciplinary Business Research*, 3(3), 422–432. https://www.researchgate.net/publication/321746840_Risk_Management_in_Universities [Accessed December 16 2020].

Association of Governing Boards of Universities and Colleges (AGB) and United Educator (UE). (2009). *The state of enterprise risk management at colleges and universities today.* https://www.vsu.edu/files/docs/internal-audit/the-state-of-enterprise-risk-manangement-at-colleges-and-universities-today.pdf

Barnabe, F., Riccaboni, A. (2007). Which role for performance measurement systems in higher education? Focus on quality assurance in Italy. *Studies in Educational Evaluation*, 33, 302–319.

Bialik, K. (2019). Why you need a data audit and how to conduct it. https://blog.capterra.com/how-to-conduct-a-data-audit/

Bozkus Kahyaoglu, S. (2020). Internal Audit Adding Value to the Organization, Chapter title: Assurance Model Approach as a Numerical Method in Internal Auditing, Seçkin Publishing House, Editor: Halis Kıral, Number of Editions: 2, Number of Pages 528, in Turkish. ISBN:9789750259555.

Bozkus Kahyaoglu, S., Sarıkaya, R., Topal, B. (2020). Continuous auditing as a strategic tool in public sector internal audit: The Turkish case. *Journal of Selçuk University Social Sciences Vocational School* (Selçuk Üniversitesi Sosyal Bilimler Meslek Yüksekokulu Dergisi), 23(1), 208–225. DOI: 10.29249/selcuksbmyd.670261

Brewer, A., Walker, I. (2011). Risk management in a university environment. *Journal of Business Continuity and Emergency Planning*, 5(2), 161–172.

Broadbent J. (2007). Performance measurement system in and of Higher education institutions in England: Professionalism, managerialism and management. *Roehampton Research Paper*. http://rrp.roehampton.ac.uk/bsspapers/3

Brown-Liburd, H., Vasarhelyi, M. A. (2015). Big data and audit evidence. *Journal of Emerging Technologies in Accounting*, 12(1), 1–16.

Burch, S. (2011). Building an internal audit function. *Internal Auditor*. http://www.theiia.org/intauditor

Cangemi, M. P. (2015). Staying a step ahead internal audit's use of technology. http://raw.rutgers.edu/docs/wcars/35wcars/Presentations/Cangemi-IIA%20Format%20IA%20Use%20of%20Tech%20July%202015%20-%20Rutgers%2011-6-15%20FINAL%20%20version.pdf

Chan, D. Y., Chiu, V., Vasarhelyi, M. A. (2018). *Continuous auditing: A book of theory and application*. Bradford, UK: Emerald Group Publishing Limited.

Clyde-Smith, J. (2014). Utilizing enterprise risk management strategies to develop a governance and operations framework for a new research complex: A case study. *Journal of Higher Education Policy and Management*, 36(3), 327–337.

Cohen, M., Rozario, A., Zhang, C. A. (2019). Exploring the use of robotic process automation (RPA) in substantive audit procedures. *The CPA Journal*, 89(7), 49–53.

COSO. (2019). *Managing cyber risk in a digital age*. https://www.coso.org/Documents/COSO-Deloitte-Managing-Cyber-Risk-in-a-Digital-Age.pdf

D'Onza, G., De Santis, F. (2020). Professional impact of digital audit – Thinking beyond the realm of audit standards. *Journal of EU Court of Auditors*, 1, 141–145. Big Data & Digital Audit. No. 1. 2020.

De Santis, F. (2016). Auditing standard change and auditors' everyday practice: A field study. *International Business Research*, 9(12), 41. DOI: 10.5539/ibr.v9n12p41

Deloitte. (2019). *Government jobs of the future. What will government work look like in 2025 and beyond?* https://www2.deloitte.com/content/dam/insights/us/articles/4767_FoW-in-govt/DI_Algorithm-auditor.pdf

Edwards, F. (2012). The evidence for a risk-based approach to Australian higher education regulation and quality assurance. *Journal of Higher Education Policy and Management*, 34(3), 295–307.

Eslami, M., Vaccaro, K., Karahalios, K., Hamilton, K. (2017). *Be careful; things can be worse than they appear: Understanding biased algorithms and users' behavior around them in rating platforms*, pp. 62–71. Montreal, Canada: Association for Artificial Intelligence (AAAI) Press.

European Commission/EACEA/Eurydice. (2020). *The European higher education area in 2020: Bologna process implementation report*. Luxembourg: Publications Office of the European Union. EC-02-20-828-EN-N. ISBN:978-92-9484-356-2. DOI: 10.2797/756192

Galloway, D. (2010). *Internal auditing: A guide for the new auditor*. Altamonte Springs, FL: The Institute of Internal Auditors Research Foundation.

Haynes, P., Ip, K., Saintas, P. et al. (2004) Responding to technological change: IT skills and the academic teaching profession. *Active Learning in Higher Education*, 5(2), 152–165.

Holt, J. E. (2012). A high-performing audit function. *Internal Auditor*. http://www.theiia.org/intauditor

IIA. (2019a). Relationships of trust building better connections between the audit committee and internal audit. *IIA Position paper*. https://na.theiia.org/about-ia/PublicDocuments/Relationships-of-Trust.pdf

IIA. (2019b). The internal audit charter a blueprint to assurance success. *IIA Position Paper*. https://na.theiia.org/about-ia/PublicDocuments/PP-The-Internal-Audit-Charter.pdf

IIA CBOK. (2015). 2015 Global pulse of internal audit. *Embracing opportunities in a dynamic environment*. https://www.aiiaweb.it/sites/default/files/imce/pdf/global_pulse_final_pdf.pdf

Kageyama A. (2014). The implementation process of enterprise risk management in higher education institutions. *International Review of Business* No. 14, 61–80. https://core.ac.uk/download/pdf/143635076.pdf

Kahyaoglu Bozkus, S. (2019). An analysis on the implementation of new approaches and techniques in the auditing of business processes based on Blockchain technologies, Chapter 6. In B. Darici & F. Ayhan (Eds.), *Cryptocurrencies in all aspects*. New York: Peter Lang Publications. ISBN:978-3-631-78387-0.

Knechel, W. R. (2013). Do auditing standards matter? *Current Issues in Auditing*, 7(2), 1–16. DOI: 10.2308/ciia-50499

Kokina, J., Davenport, T. H. (2017). The emergence of artificial intelligence: How automation is changing auditing. *Journal of Emerging Technologies in Accounting*, 14(1), 115–122. DOI: 10.2308/jeta-51730

KPMG (2020). *20 key risks to consider by Internal Audit before 2020 Are you aware of the risks concerning Internal Audit today and in the near future?* https://assets.kpmg/content/dam/kpmg/ch/pdf/key-risks-internal-audit-2018.pdf

Lundquist, A. E. (2015). Lessons from the academy: ERM implementation in university settings, Chapter 9. In John R.S. Fraser, Betty J. Simkins, and Kristina Narvaez (Eds.), *Implementing enterprise risk management* (pp. 143–178). Hoboken, NJ: John Wiley & Son, Inc.

Martino, P., D'onza, G., Melville, R. (2019). The relationship between CAE leadership and the IAF's involvement in corporate governance. *Journal of Accounting, Auditing & Finance*, 36(2), 1–19. DOI: 10.1177/0148558X19867539

McCaffery, P. (2019). *The higher education manager's handbook effective leadership and management in universities and colleges*, 3rd ed. London, UK: Taylor and Francis Publications. ISBN:978-1-351-24974-4 (ebk).

Moffitt, K. C., Rozario, A. M., Vasarhelyi, M. A. (2018). Robotic process automation for auditing. *Journal of Emerging Technologies in Accounting*, 15(1), 1–10.

NACUBO and AGB. (2007). *Meeting the challenges of enterprise risk management in higher education*. https://www.semanticscholar.org/paper/Meeting-the-Challenges-of-Enterprise-Risk-in-Higher-Mattie/054cf1a44b14a2cac449212d8947356ffcd6eaea

Pickett Spencer, K. H. (2012). *The essential guide to internal auditing*, 2nd ed. West Sussex, England: John Wiley & Sons.

Pitt, S. A. (2014). *Internal audit quality developing a quality assurance and improvement program*. New Jersey, USA: Wiley Publications. ISBN:978-1-118-71550-5 (ePDF).

Ramamoorti, S., Siegfried, A. N. (2016). Promoting and supporting effective organizational governance. Internal Audit's Role. The Institute of Internal Auditors Research Foundation (IIARF). ID #2016-0423.

Rezaee, Z. (1996). Improving the quality of internal audit functions through total quality management. *Managerial Auditing Journal* 11(1): 30–34.

Rezaee, Z. (2010). The importance of audit opinions. *Internal Auditor*. http://www.theiia.org/intauditor

Rozario, A. M., Thomas, C. (2019). Reengineering the audit with blockchain and smart contracts. *Journal of Emerging Technologies in Accounting*, 16(1), 21–35.

Ruzic-Dimitrijevic, L., Dakic, J. (2014). The risk management in higher education institutions. *Online Journal of Applied Knowledge Management, International Institute for Applied Knowledge Management*, 2, 137–152. https://www.researchgate.net/publication/321746840_Risk_Management_in_Universities [accessed December 16 2020].

Salijeni, G., Samsonova-Taddei, A., Turley, S. (2019). Big Data and changes in audit technology: contemplating a research agenda. *Accounting and Business Research*, 49(1), 95–119.

Sandvig, C., Hamilton, K., Karahalios, K., Langbort, C. (2014). Auditing algorithms: Research methods for detecting discrimination on internet platforms.

Schreckling, E., Steiger, C. (2017). Digitalize or drown. In G. Oswald & M. Kleinemeier (Eds.), *Shaping the digital enterprise*. Switzerland: Springer International Publishing. DOI: 10.1007/978-3-319-40967-2_1

Soares, L., Steele, P., Wayt L. (2016). *Evolving higher education business models: Leading with data to deliver results*. Washington, DC: American Council on Education.

Tang, J., Karim, K. (2017). *Big data in business analytics: Implications for the audit profession*. Available at: https://www.cpajournal.com/2017/06/26/big-data-business-analytics-implications-audit-profession/

Tufano, P. (2011). Managing risk in higher education. *Forum for the Future of Higher Education*, 54–58.

URMIA. (2007). *ERM in higher education*. Bloomington, USA: University Risk Management and Insurance Association (URMIA).

Vasarhelyi, M. A., Cho, S., Cheong, A., Chanyuan (Abigail), Z. (2020). Smart audit: The digital transformation of audit. *Journal of EU Court of Auditors*, 1, 27–32. Big Data & Digital Audit. No. 1. 2020.

Wang, X. (2010). Performance measurement in universities. Managerial Perspective. Business Administration – Financial Management Faculty of Management and Governance University of Twente. Supervisors: Prof. dr. Nico P. Mol and Dr. Ben Jongbloed.

Yilmaz, K. (2013). Comparison of qualitative and quantitative research traditions. Epistemological, theoretical, and methodological differences. *European Journal of Education*, 48(2), 311–325.

Yoon, K., Hoogduin, L., Zhang, L. (2015). Big data as complementary audit evidence. *Accounting Horizaons*, 29(2), 431–438.

Zhang, C. (2019). Intelligent process automation in audit. *Journal of Emerging Technologies in Accounting* 16 (2), 69–88.

Zhang, J., Yang, X., Appelbaum, D. (2015). Toward effective big data analysis in continuous auditing. *Accounting Horizons*, 29(2), 469–476.

Chapter 5

Internal audit resourcing and staffing based on digitally enriched learning spaces

5.1 DETERMINING THE DIGITAL SKILLS FOR INTERNAL AUDIT STAFFING AND RESOURCING

The main idea of this book is to propose that internal auditors have a strategic role as "trusted advisors" in managing the new digital risks faced by universities and the challenges of the digital transformation process. If internal auditors do not have this strategic perspective, it will be impossible for them to fulfill the role expected from them, even causing the organization they serve to fall behind in achieving the goals. Therefore, in this chapter, the digital competencies of internal auditors are explained in detail regarding internal audit staffing and resourcing. In this context, it is preferred to discuss the requirements of digital skills in terms of the realities, expectations, and strategic priorities of the higher education sectors.

It is important for internal audits to carefully monitor trends in the HE sectors and to plan resources and budgets based on the strategic direction of the institution. It should be noted here that online or blended learning can change the nature of a higher education experience and indeed the nature of higher education institutions. In this context, there may be some key questions:

- Will campuses really continue to exist as the main business model for organizing and delivering higher education in a digital transformation environment?
- In this context, will many facilities in higher education institutions – accommodation, sports facilities, social services, etc., continue to be an integral part of higher education?
- How will digital learning and teaching impact public and private funding models, including student support?
- How will be learning and teaching environment relate to the digital environment?
- How will digital technologies contribute to the higher education policy objectives?

DOI: 10.1201/9781003093008-5

Internal audit needs to reveal "value proposition" when seeking answers to these questions contained here in the context of their terms of reference and responsibilities within the HE institution.

According to European Commission's Education, Audiovisual and Culture Executive Agency (EACEA) (2020), it is very important to understand digitalization problems in the context of equipping individuals for "lifelong learning" in a rapidly changing technological environment. Therefore, EC – EACEA (2020) claims that higher education institutions should see this need and first review their business models to reach large groups in society and provide the necessary support.

It is a fact that digital tools in higher education have great potential in terms of both providing quality education and reinforcing social relations. However, it should be noted that this development will not occur automatically. In other words, benefiting from digital technology in line with appropriate corporate strategies will require broad reflection on various issues (Bonnet and Nandan 2011). In this context, strategic policy planning has an important role for both HE institutions and the HE sectors.

Based on the strategic directions and business model of the higher education sectors, the CAEs should determine their strategic plan for digital transformation in line with the vision of the HE institution. Accordingly, while CAEs define the future resources and competencies of internal audit, institutional requirements, and stakeholder expectations should be decisive. A summary representation of this is included in the figure. Hence, CAE should have to find an answer to the questions in Figure 5.1 to establish the

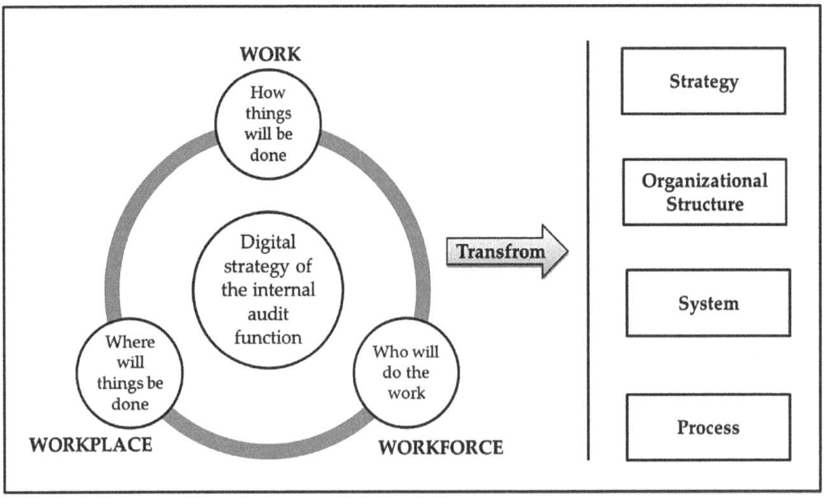

Figure 5.1 Key elements of digital strategy of internal audit function.

Source: Adapted from Deloitte (2019).

internal audit function in line with the strategic goals of the university. These questions are directly related to defining "the work", "the workforce", and "the workplace" of the internal auditors. In this way, necessary recruitment can be made to reach out to the predefined strategy, organizational structure, system, and process of the internal audit function.

The IIA (2020) conducted a study to define internal auditor competencies globally and raise awareness of the recommended competencies updating them according to innovation and technological needs. In this newly published competency matrix study on the internal auditors, the types of competencies needed to successfully perform and maintain each stage of internal audit are explained and measured in detail.

According to IIA's Internal Audit Professional Practice Standards, the chief audit executive must establish a program for identifying, selecting, and developing the human resources required for an internal audit department (Standard 540). Within the scope of the program in question, the human resources planning, and development process of the internal audit department should be carried out within the following framework:

- A written job description should be created for all personnel working in the internal audit department. In particular, it would be appropriate for this definition to include the work program required in the digital transformation strategy (Figure 5.1). The most important resource is considered as human resources to be able to master the complex business processes of universities and to reach the required level of assurance. Therefore, the information included in the job descriptions must be compatible with the corporate development process and relevant for the current HE organizational structure and business model
- Efforts should be made to ensure that those who work in the internal audit department are qualified and competent. It is important that the chief audit executive makes a decision in coordination with the audit committee and the board of directors and allocates a sufficient budget for which qualifications and competencies are achieved
- Training and continuing education opportunities must be provided to each internal auditor. For its continuity, it is important that the audit culture is open to training and support
- It is necessary to evaluate the individual performance of each internal auditor assigned in the internal audit department and their fitness with the team at least annually
- It will be beneficial for the chief audit executive to provide consultancy to the internal auditors on their performance and professional development when deemed necessary (Nzechukwu 2017)

When the literature is reviewed, it is seen that competence is used frequently and various definitions are given in practice. According to Weinert (1999), the concept of competence is expressed as a combination of the preconditions that individuals are expected to possess, for instance, the abilities, capacities, and skills required to achieve certain goals.

Among these, performance-related competency indicators contain more features than other criteria. This situation cannot be considered a coincidence (Desmet et al. 2015). At the same time, when the content is examined closely, it is seen that emphasis is placed on data analysis with the use of advanced technological tools and techniques. Thus, recommendations are provided regarding what should be the digital competencies of auditors in the face of increasing digital audit and risk universe. CAE should consider these recommendations to achieve best practices regarding resourcing and staffing their IA departments.

According to SAP (2015), successful organizations are particularly "agile" in three main areas: rethinking business models, rethinking business processes, and rethinking work. It should be noted that when dealing with digital capabilities, it is necessary to analyze this largely specific to the context of the organization concerned. In particular, the digitalization strategy should be evaluated as "institution-specific" and an in-depth analysis should be made based on the socio-economic conditions of the institution. The main reason for this is that a strategy that works and brings success in one organization may not work well in another (vom Brocke and Rosemann 2014; vom Brocke et al. 2016). Accordingly, aligning the dynamics of the higher education sector as a global market on the one hand and strategically positioning the individual DNA of a university, on the other hand, plays an important role.

With the effect of digitalization, there are two main issues that internal auditors should take into consideration. The first is a change in the way of working in the higher education sectors; the second is the increase in the speed of change faced by universities significantly. Considering both issues, three main requirements must be addressed for a higher education institution to be successful as follows: "New skills and competencies", "new leadership styles", and "new institutional skills". Therefore, depending on the degree to which universities meet these three requirements, it may be possible for them to develop their culture toward a "digital mindset" and mature the level of digital culture (Kohnke 2017).

However, digital transformation initiatives generally require challenges to realize the expected benefits. In current research and literature, the importance of managing people and organizational problems in digital transformations is clearly shown based on the practices of different countries (Bonnet and Nandan 2011; Fitzgerald et al. 2013; McAfee and Welch 2013; Bughin et al. 2014; Westerman et al. 2014; Bughin et al. 2015; Kohnke 2017; Kandeh and Alsahli 2020).

Organizational change management is a critical success factor for a digital transformation program. Accordingly, while implementing change management, actions should be taken in four main areas. These are respectively: "Aligning leadership" by promoting the digital vision and role modeling, "mobilizing the organization" by supporting open communication, "developing skills" by providing resources to gain digital skills, and "maintaining sustainability" by defining KPI and adaptation of incentive systems (Kohnke 2017). The summary of the change management process of digital transformation in the HE sectors is shown in Figure 5.2.

Due to the structure of higher education institutions, there may be diversity among stakeholders in some cases in the value creation process and even operate in a "multi-stakeholder" context where there may be conflicting expectations between stakeholders (Lääts et al. 2019).

At the executive level, these contradictions must be eliminated, managed properly. And the balance between interests must be pursued. The views of all stakeholders should be considered when making benefit, resource, and risk-assessment decisions at the management level. Lääts et al. (2019) from the Cooperation in Higher Education for Digital Transformation in European Business (CHEDTEB), recommends that the following be questioned regarding each decision:

- Who benefits?
- Who bears the risk?
- What resources are needed to achieve strategic goals?

The decisive parameter is to determine the strategic direction of a university, considering the needs of stakeholders. Management generally ensures that its strategic goals are achieved by evaluating the needs, status, and preferences of stakeholders. In addition, management determines strategic direction through prioritization and decision-making process and monitors

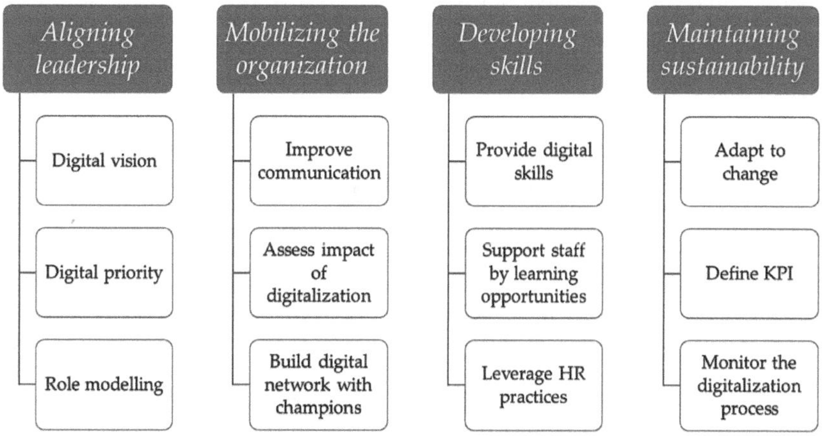

Figure 5.2 Change management process of digital transformation in the HE sectors.

Source: Adapted from Kohnke (2017).

performance, compliance, and progress according to agreed directions and goals of the HE institution (ISACA-COBIT 5 2012; Garsoux 2013). Higher education institutions may prefer different solutions and standardization levels that are given as follows:

- "Server-based decentralized": this level includes digital solutions specific for faculties
- "Server-based or web-based centralized": this level indicates solutions for a university
- "Web-based information sharing": this level indicates collaboration between universities and other institutions

In this context, the role of internal auditors changes based on the level of digitalization and standardization levels. In addition, CAEs should plan their staffing and make resourcing to achieve their roles and responsibilities accordingly (Lääts et al. 2019; Bozkus Kahyaoglu and Coskun 2020). The CAE's special focus should be on reviewing the roles of the internal audit department to execute the skills and resource planning process. These roles can be covered in three main levels. "Strategic role", "legal role", and "operational role", respectively (Figure 5.3).

Accordingly, the necessary coordination should be established with the Audit Committee regarding the strategic role of internal audit, and the competency requirements and resource planning of the internal audit management team should be made to achieve the strategic goals and responsibilities (KPMG 2017).

While fulfilling its legal role, it is necessary to consider the higher education sector in which universities operate and the regional, national, and international requirements relevant for the HE institution. In addition, it is important to determine the resource planning and competencies by taking into account the legal regulations for monitoring compliance with HE

Figure 5.3 Major roles of internal audit in the university.

Source: KPMG (2017, 2019).

quality and accreditation standards in higher education institutions with an international perspective (KPMG 2019).

On the other hand, resource planning should be done based on its role in auditing the operational works and processes encountered in every university regardless of its scale. It would be more appropriate to perform the operational role of audit based on automatic and remote monitoring with technology support. For this, it may be possible in the future to employ robotic auditor applications with technological investment instead of recruiting human internal auditors (Chartered Institute of Internal Auditors 2019).

5.2 DIGITAL REQUIREMENTS FOR AN EFFECTIVE INTERNAL AUDIT

While predictions are made with the future of work in the literature, predictions are also made for prospective auditors. First, it will not be sufficient for future auditors to be specialized only in accounting, finance, and compliance. As a critical requirement of data analysis, programming, coding, and mathematical knowledge are gaining importance (Kandeh and Alsahli 2020). Accordingly, the basic views and comments on how the auditors will be in the future are summarized below.

Considering the audit and risk control environment, digitization is a determining issue to understand the main drivers and abilities by using such digital technology breakthroughs like "(s)ocial media", "(m)obile computing", "(a)nalytics/big data", "(c)loud computing". These are called with an integrated view as "SMAC". In addition, "the Internet of Things (IoT)", "cyber-physical systems (CPS)", "cyber-human systems (CHS)", and "cyber-security systems (CSS)" can be added to the long list (SAP 2015; Kowalkiewicz et al. 2016).

It is seen that digital technologies have a wide reach in the higher education sector and are becoming widespread quite rapidly. It is stated that the underlying reason for the rapid spread of digitalization are "Moore's law"[1] and "Metcalf's law".[2]

Processes and workflows have long been objects of digital transformation in the HE sectors. The main difference here is that in the past when it came to digitalization, it only concerned data management and processing in the IT departments of the HE institutions. However, it is now in a position that radically affects all departments of the HE institutions. Therefore, when it comes to digitalization, today's meaning covers everything from management and support tasks to core business processes in the HE sectors (Kane et al. 2015). Considering today's pressures and competitive environment, universities are expected to go far beyond automating their current processes. To meet increasing social expectations, it has often become a necessity for them to reinvent all business processes (Markovitch and Willmott 2014).

As a result of digitalization in audit, data processing will become more important. In this respect, especially in the audit fieldwork processes in the digital environment, the internal auditor should have a new tool considering the process of handling, analyzing, and reporting data, as well as operational processes. In this sense, it becomes a necessity for an internal auditor to have the necessary competency in statistics, mathematics, and computer programming as well as auditing standards (Bozkus Kahyaoglu and Coskun 2020; Kandeh and Alsahli 2020). This interdisciplinary approach in digital audit fieldwork will highlight ethical standards beyond the IPPF standards in audit processes. For this reason, the establishment, control, and compliance of ethical standards in the model, data analysis process, and reporting of data will become the most influential business processes of the digital audit in the new era (Kokina and Davenport 2017).

The digitalization trend in the higher education sector today brings great opportunities, challenges, or both for all HE institutions. Thus, digital talents are increasingly in a position to be able to determine whether a university is creating value or not (Hirt and Willmott 2014). Based on this fact, internal auditors need to review their digital competencies to align with the stakeholder expectations.

The principal point is that the competencies of the auditors are compatible with the audit objectives and scopes (IIA 2020; Kandeh and Alsahli 2020). Instead of all internal auditors having competencies in the equivalent areas, it would be appropriate to create an audit team with expertise in different realms. The main reason for this is that the internal audit team should have a range of competencies that encompass the entire audit universe because of the different types of digital risks and advanced technological information encompassed by the digitalized business model of higher education institutions. It should not be forgotten that outsourcing and/or co-sourcing is possible in areas of expertise where resources are inadequate. It is important to avoid approaches that may cause "audit risk" as an outcome of the evaluation of the competencies of internal auditors as being biased or overly confident (Bozkus Kahyaoglu 2020).

When it comes to digitalization, even if it is generally perceived as a single field, this perception can be quite misleading from an internal audit perspective (Bozkus Kahyaoglu and Coskun 2020). This is because a university subject to digital transformation is exposed to various risks and challenges within an audit perspective. It should be noted that with the advancement in technology, digital risk types are diversifying very fast. On the one hand, the need for the analysis of rapidly increasing big data arises, on the other hand, the need to analyze the control weaknesses arising from the use of advanced technology as a tool in IT infrastructure and HE sector core business processes. This makes it necessary for auditors to have very high digital competencies and to acquire deep knowledge in different IT-related subtopics such as data verification, data analytics, AI algorithms, cybersecurity, business continuity, authentication, blockchain, robotics, etc. In this context, it is recommended that internal auditors take a systematic approach to cope with data analytics, AI, and cybersecurity issues and develop their competencies accordingly (Brown-Liburd and Vasarhelyi 2015; Kokina and Davenport 2017).

5.3 DEVELOPING DIGITAL SKILLS AND KNOWLEDGE ON DIGITALLY ENRICHED LEARNING SPACES WITHIN INTERNAL AUDIT

According to Salijeni et al. (2019), internal auditors cannot maintain a working style in which they only report their findings in the global competitive environment and under the influence of technological developments. Instead, it is recommended to work as a knowledgeable in-house advisor who develops further recommendations on effective governance, risk management, compliance, and quality, making constructive suggestions to line managers on how to improve business performance in the digitally transforming environment. In this context, internal audit and internal control environments must have more attention and resources required to take action against increasing types of digital risk (COSO 2020).

In a time of rapidly increasing digital risks in the HE sectors, internal auditors need to revise their digital capabilities to recognize, understand, analyze them, and provide recommendations that add value to the HE institution.

According to the research findings made by Protiviti (2020a), it is emphasized that among the features and competencies that internal auditors should have in the future, there should be "corporate competencies" as well as "digital competencies". That increases in-house positioning, strong communication and the effectiveness of knowledge sharing, and more cooperation and interaction with those involved in the business processes in teamwork. Thus, it will ensure that future internal auditors have real-time productivity within the framework of the smart audit strategy.

According to Protiviti's (2020a) next-generation internal auditor research report, internal auditors are already making their way in many leading organizations. Here, the value created by internal auditors reflects a strategic perspective that emphasizes the foresight of the future rather than a perspective that provides solutions only to the present challenges. For this, as Albert Einstein said, "it is necessary to start by changing our mindset first". That is an issue that is expected to be at the top of the chief audit executive's agenda. So, the adaptation is starting from changing the mentality to change the behavior.

From the strategic point of view of the next generation of internal auditors, there are long-term projections, sustainability, and growth issues relevant for their organization. It is imperative to benefit from advanced technological tools and methods in obtaining this foresight and providing evidence-based recommendations (Suffield 2020; Protiviti 2020a, 2020b). Therefore, the expectation of an auditor in the future would be acting ahead of the process owners. He/she would guide them persuasively with his thoughts and behaviors.

Vasarhelyi et al. (2020) describe the audit in this new era as "the age of smart audit". Therefore, a new audit culture is emerging on the agenda of smart auditors, supported by analysis of big data and advanced audit analytics tools. The vital point here is that the data that process owners are constantly producing is growing. However, the number of auditors increases much is neither in any part of the world nor in any other sectors. As a general tendency, while the number of auditors is diminishing, there is an increased investment in technological infrastructure, facilities, and audit tools.

To keep up with this age of smart auditing, an internal auditor must know how to examine data and what to look for to achieve strategic audit objectives. Accordingly, auditors are offered more tools, namely audit analytics, to examine big data. In addition, it is possible to sort out routine tasks through "automation tools" such as "Robotic Process Automation (RPA)". Thus, the hands of internal audit are freer to be more creative. These types of audit tools not only make auditors smarter but also enable them to focus on more productive tasks and do value-added work (Protiviti 2020a).

Considering technological developments and innovations, the first things that come to mind are artificial intelligence, deep learning, and robotic applications. However, this ranking has not yet been fully reflected in the internal auditors' agenda. In this context, the findings of the research report prepared by Protiviti (2020a) are reviewed. When the changes in response between 2019 and 2020 the concept of "enabling technology competencies – being undertaken – to transform audit process"; it was determined that there was a downward trend rather than a forward-looking development trend.

The main reason is that the digital competencies of internal auditors are not yet fully mature, as well as the slowdown in the pace of institutions investing in such new technologies and the decline in overall growth rates in the economy worldwide due to the pandemic period. That is given in Figure 5.4 and hence, looking closely at Figure 5.4, it can be interpreted as follows. Protiviti's (2020a) highest self-assessment score for internal audit digital maturity levels included in the appendix of the same research report is 7 out of 10. This means that digital change and innovative transformation do not seem to be resolved in such a short time. On the contrary, it makes it necessary to reflect the philosophy of "lifelong learning" to work with a long-term approach that requires great patience for internal auditors.

The key point here in Figure 5.4 is that internal auditors concentrate on advanced audit analysis techniques. This is a good indicator of their approach to digitalization since most audit package programs already contain many different AI, ML, and DL analysis methods in their background and algorithms. Even if the internal auditors do not directly analyze the big

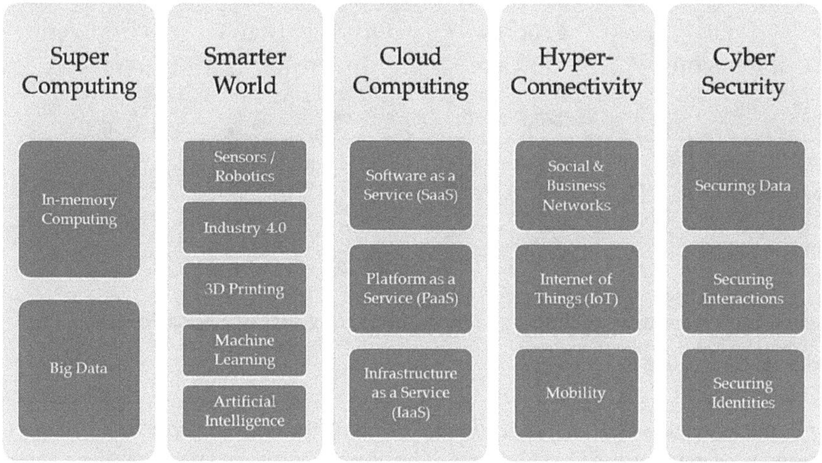

Figure 5.4 Enabling technology competencies being undertaken to transform audit process.

Source: Protiviti (2020a).

data themselves, it can be stated that they are currently acumen in using advanced audit tools and communicating the findings obtained with it.

5.4 CHALLENGES INTERNAL AUDITORS FACE WHEN EXECUTING DIGITAL AUDIT OBJECTIVITY AND PROFESSIONAL SKEPTICISM

It is a fact that when the core values and priorities of internal auditors are aligned with the strategic values of the organization in which they serve, their working relationship will inherently become more coherent. Thus, decision-making processes are straightforward and communication styles will be similar. Conversely, a stressful experience can arise when an internal auditor's core values do not fully align with the organization.

The existence of change and technological developments can be expressed as the discourse of every period. In this respect, having the ability to adapt to change will provide the strength and perspective needed to deal with challenges that may be encountered. The crucial primary point here is that a revolutionary period has entered the innovation and digital transformation process. Breakthrough new technologies have matured to a certain extent and made five key technological advances widespread and available (SAP 2015; Oswald and Kleinemeier 2017). These five major trends impacting the higher education sector are summarized in Figure 5.5.

When considering the digital transformation from an audit perspective, the ability to understand performance criteria, comparison of organizational

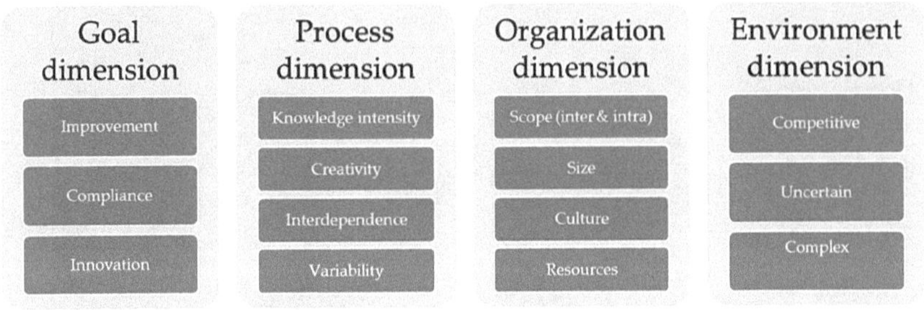

Figure 5.5 Five major technology trends impacting higher education sectors.

Source: SAP (2015).

or process results with this standard, and identifying and reporting the deviations is of paramount importance. That approach is at the core of an internal auditor's critical competencies and is the defining feature for an "effective auditor".

It is a fact that the main goal of university management is to increase performance by digital transformation. However, there may be universities that do this in practice and others that fail to do so. Two basic concepts are emphasized while defining performance in the literature. The first is efficiency and the second is effectiveness. Drucker (1994, 2008) expresses efficiency as "doing things right". For universities, this means controlling the costs incurred in a particular business process and, if possible, lowering it and thus increasing productivity. The second concept, effectiveness, is expressed as achieving goals that provide an advantage over peers (Barney 1991). The key point internal auditors should consider here is that, unlike efficiency, effectiveness also includes the action of "doing the right thing".

In the literature, there are different views and options regarding the situations encountered in digital transformation. On the one hand, there are business process owners who try to gain a competitive advantage by innovating in business processes; on the other hand, there are IT process owners who aim to ensure business continuity. If there is harmony and cooperation between them, positive effects of agility and digitalization on corporate performance can be achieved. In this context, internal audit has an extensive role in ensuring "agility", "flexibility", and "sustainability" through this "harmony" and cooperation throughout the organization (Kharabe and Lyytinen 2012).

According to Butera (2016), changes in an internal audit department are a problematic issue. The main reason for this is the behavioral preferences and characteristics of internal auditors. The main and most important of these can be said as "critical thinking" ability. This means greater for internal auditors. Because an internal auditor with critical thinking ability means being able to develop strategic ideas, evaluate the opinions of others, and reflect on internal audit fieldwork by maintaining professional skepticism. In this respect, it never means finding fault by condemning (Butera 2016).

Critical thinking is essential among the performance factors of internal audits. Critical thinking requires, first and foremost, good assimilation of knowledge. It is also the art and science of analyzing that information, synthesizing, and evaluating for the need to gather audit evidence (Wang 2010). Following this analysis and synthesis, the time comes to seek and determine what course of action is required for a clear understanding of root risk, control, or lack. By addressing the issue with a broad perspective associated with it, it aims to identify whether a related problem or additional information is required or not (Butera 2016; Kandeh and Alsahli 2020).

On the other hand, if the defining characteristic discussed here is overused by an internal auditor, it can cause the auditor to become rigid. That may cause the auditor in question to fail to detect gray areas in a business situation encountered during fieldwork. Similarly, when this preference is overused by an internal auditor, the auditor becomes increasingly dependent on and relies on checklists. Thus, the internal auditor starts to think less and less critically. As this situation becomes permanent, audit scopes and objectives do not reflect the specific aspects of the organization. Even the audit reports prepared by this type of auditor may start to reflect repetitive information.

In addition to financial data arising from accounting and financial transactions, the analysis of non-financial data is now gaining importance in the HE sectors. This situation requires internal auditors to learn how to analyze non-financial data and financial data and use it as a tool in the audit fieldwork (SAP 2015). The most significant example of this is that internal auditors obtain the opportunity to detect possible fraud cases early, as a result of the findings obtained were based on non-financial data, especially during the pandemic period (Kandeh and Alsahli 2020; Vasarhelyi et al. 2020).

Vasarhelyi et al. (2020) state that the nature of data is changing in digital speed and big data has become more interlinked to various sources. The authors call this "data ecosystems". This situation leads to difficulty in adjusting the accounting and auditing profession and standards as well. In this context, the authors argue that there is a need for revolutionary change and transformation in accounting science and audit approaches. The frequency foreseen for the assurance processes and financial statements is gradually shortening and rapidly evolving into real-time monitoring and auditing models. Also, the impact of blockchain technologies is revealing irreversible changes in both accounting and auditing (Bozkus Kahyaoglu 2019; Bozkus Kahyaoglu et al. 2020).

5.5 KEYS TO EXECUTING INDEPENDENCE, OBJECTIVITY, AND PROFESSIONAL SKEPTICISM IN DIGITALLY ENRICHED LEARNING SPACES

Today, internal auditors are expected to analyze the rapidly increasing big data in HE institutions faster. At the same time, it is now expected to perform the analysis of the population instead of the audit sample, and to produce findings based on its interpretation, make predictions, and provide value-added recommendations. In this context, according to Brown-Liburd and Vasarhelyi (2015), it is stated that in some cases, big data harms the professional judgment of internal auditors. The main reason for this is that auditors can easily get lost in the details of big data, especially if they do not have a good plan of what to do and what data they should have.

Therefore, it is necessary to be careful when handling and auditing big data. Auditors may have to deal with gigantic amounts of big data from different sources across the departments of HE institutions. In such a situation, it is necessary to verify the data and show professional skepticism in professional judgment and decision-making regarding the data (IIA 2020). It can be misleading to directly analyze any data received from the system. Therefore, if the auditors do not have relevant information about collecting and analyzing such data within an audit task, they should produce solutions with expert support.

Bührig et al. (2018) discuss the independence, objectivity, and professional skepticism approaches that internal auditors should follow as trusted advisors in a university experiencing the digital transformation process in four main dimensions. These are referred to as "goal dimension", "process dimension", "organization dimension", and "environment dimension" which are shown in Figure 5.6.

With digital transformation, universities generally aim to improve the quality of education by increasing the effectiveness and efficiency of their main processes, managing and securing data, standardizing business processes in all campus locations including the virtual environment, and guiding them on best practices regarding campus management.

Since these objectives are tried to be achieved by encountering a resource (budget) constraint, internal auditors need to determine the level of goal realization from a "cost–benefit analysis" perspective. In addition, extending

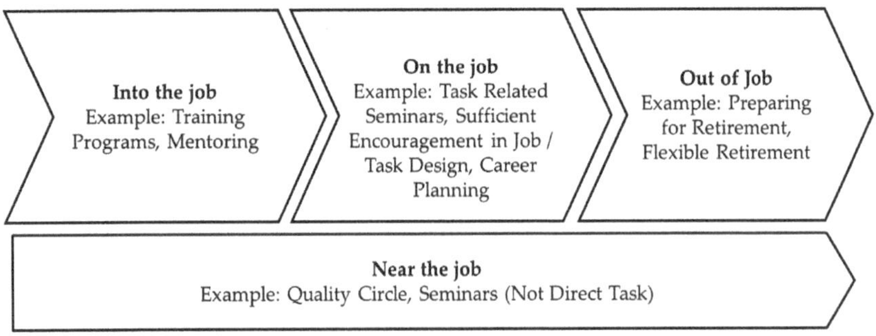

Figure 5.6 Contextual dimensions of digital transformation process in the university.

Source: Adapted from Bührig et al. 2018.

the digital transformation process longer than planned can multiply these costs. It can also lead to "reputational risk" which may consequently impact the level of strategic goal achievement. Therefore, internal auditors need to monitor the digital transformation process closely as an "observer" to reduce project delays caused by the development of new tasks.

Secondly, the business processes of universities are quite complex. Therefore, there may be variability and internal dependencies between business processes. This may cause disruptions and a slowdown in the digital transformation process. The information density that varies according to the nature and purpose of the process should also be considered. The internal audit may have suggestions that create added value in terms of standardizing the process.

Thirdly, the internal audit should take into account the organization dimension of the university in audit fieldwork. The duties of internal auditors may vary depending on whether the domains of digital transformation are handled partially or as a whole. There may be fully digitalized departments throughout the HE institution, or units that have completed their transformation in half or have not yet begun digitalization. In this case, the approach, method, tools, and audit techniques used by internal audit while auditing a division with a digital infrastructure may change. This issue should be constantly monitored, and it is necessary to decide whether the audit planning and fieldwork will be carried out remotely or physically on site. However, while the departments that have completed the digital transformation are provided with assurance services, it may be necessary to provide advisory services in departments that have not yet transformed. Regardless of the stage of the digitalization process, effective communication needs to be in place to contribute to the development of positive organizational culture.

In the traditional understanding of auditing, an internal audit department would focus on the internal control system and control environment of the HE institution and perform a risk-based audit. However, today, it is not possible to provide sufficient assurance to the stakeholders with such fieldwork and vision. Because environmental factors are now included in the corporate risk portfolio more than in the past. Internal auditors had the opportunity to make different analyzes by easily accessing external data. Regardless of the nature and source of the data that have become open through social media and network structures with digitalization, it can now be easily acquired and analyzed by the internal auditors. Within the framework of this approach, it is important that risk-based audit planning should be more dynamic and highly sensitive, and flexible to newly added risks, mostly "real-time risks". This is briefly expressed as an "agile audit" (Protiviti 2020a).

It is a fact that training is a fundamental necessity today, as it was in the past, as an important tool to sustain the personal development and professional development of the internal audit team. However, there are differences in the form, place, and duration of training. It is important that this is

2019	Current Situation	Plan for Next Year
AI/Machine Learning	%17	%32
Process Mining	%20	%27
Robotic Process Automation	%19	%25
Advanced Analytics	%25	%29

2020	Current Situation		Plan for Next Year	
AI/Machine Learning	%7	↓	%15	↓
Process Mining	%10	↓	%21	↓
Robotic Process Automation	%12	↓	%17	↓
Advanced Analytics	%23	↓	%29	=

Figure 5.7 Types of training for internal auditors.

Source: Adapted from Naurem (2020) and Nzechukwu (2017).

internalized by the internal audit team and its sustainability is ensured by getting CAE and Management support.

Although training is necessary, if it is not suitable for its purpose, it can turn into a cost item and even a waste of time. Therefore, it is recommended to explore and apply new training techniques particularly in conformity to the internal auditor profile, experience, and organizational needs. As it can be seen in Figure 5.7, the point that should be taken into consideration is to comprehensively determine what kind of training they need and monitor them among performance criteria in the corporate scorecard.

Another point to be considered here is that the satisfaction level of internal auditors regarding the working environment should be monitored continuously. It is not possible to achieve the goals with only one-sided satisfaction in order to increase the performance level. Therefore, it is recommended that the CAE conduct a satisfaction survey for employees in the internal audit department and monitor turnover rates. It is important that the criteria determined while measuring the performance are motivated and supportive toward achieving the goals and play an encouraging role.

Otherwise, a situation may arise where "Goodhart's Law" applies. Goodhart's Law generally focuses on the determination of performance criteria. It provides information that explains the negative effects of failures encountered during the benchmark-setting phase on performance. According to Goodhart (1975), if a benchmark moves away from performance improvement and becomes just a goal, it may lead to a deviation from the true goal by abuse or misuse of these criteria. Therefore, performance criteria should be applied as a tool, not an outcome for individuals to prevent any misuse and the deleterious effects.

NOTES

1 Moore's law is defined as the observation that the number of transistors in a dense integrated circuit doubles approximately every two years; see https://en.wikipedia.org/wiki/Moore%27s_law.
2 Metcalf's law states that the value of a telecommunications network is proportional to the square of the number of connected users of the system; see https://en.wikipedia.org/wiki/Metcalfe%27s_law.

REFERENCES

Barney, J. (1991). Firm resources and sustained competitive advantage. *Journal of Management*, 17(1), 99–120.
Bonnet, D., Nandan, P. (2011). *Transform to the power of digital—digital transformation as a driver of corporate performance*. Available via Capgemini Consulting. https://www.capgemini.com/resources/transform-to-the--power-of-digital
Bozkus Kahyaoglu, S. (2019). An analysis on the implementation of new approaches and techniques in the auditing of business processes based on blockchain technologies. In B. Darici & F. Ayhan (Eds.), *Cryptocurrencies in all aspects*. New York: Peterlang Publications. ISBN:978-3-631-78387-0.
Bozkus Kahyaoglu, S. (2020). Assurance model approach as a computational method in internal auditing (İç Denetimde Sayısal Yöntem Olarak Güvence Modeli Yaklaşımı). In H. Kıral (Ed.), *Internal audit adding value to the organization (İç Denetim Kuruma Değer Katmak)*, 2nd ed. Ankara, Turkey: Seçkin Publishing.
Bozkus Kahyaoglu, S., Coskun, E. (2020). *Rethinking audit culture in digital transformation era: Recommendations for auditing digitally transforming universities*. In *The 11th international conference on governance fraud ethics and corporate social responsibility (CSR) which was virtually hosted by London Metropolitan University, Guildhall School of Business and Law* on 27–28 August, 2020.
Bozkus Kahyaoglu, S., Sarıkaya, R., Topal, B. (2020). Continuous auditing as a strategic tool in public sector internal audit: The Turkish case. *Journal of Selçuk University Social Sciences Vocational School (Selçuk Üniversitesi Sosyal Bilimler Meslek Yüksekokulu Dergisi)*, 23(1), 208–225. DOI: 10.29249/selcuksbmyd.670261
Brown-Liburd, H., Vasarhelyi, M. A. (2015). Big data and audit evidence. *Journal of Emerging Technologies in Accounting*, 12(1), 1–16.
Bughin, J., Chui, M., Pollak, L. (2014) Organizing the networked enterprise for change. *McKinsey on Business Technology* (33), 50–57.
Bughin, J., Holley, A., Mellbye, A. (2015) *Cracking the digital code: McKinsey global survey results*. McKinsey & Company. Available via McKinsey & Company. http://www.mckinsey.com/business-functions/business-technology/our-insights/cracking-the-digital-code
Bührig, J., Schoormann, T., Knackstedt, R. (2018) Business process management in german institutions of higher education: The case of Jade University of Applied Science. In: vom Brocke, J., Mendling, J. (Eds.), *Business Process Management Cases. Management for Professionals*. Cham: Springer. https://doi.org/10.1007/978-3-319-58307-5_31

Butera, A. (2016). Chapter 8: Techniques for managing the constituent. In D. Swanson (Ed.), *Mastering the five tiers of audit competency: The essence of effective auditing*. Boca Raton, FL: CRC Press Taylor & Francis Group. ISBN:978-1-4987-3849-1.

Chartered Institute of Internal Auditors. (2019). Effective risk management – Joint internal audit and risk management functions. *Position Paper*. https://www.iia.org.uk/resources/risk-management/position-paper-risk-management-and-internal-audit

COSO (2020). Blockchain and Internal Control. The COSO Perspective. Sponsored by Deloitte. USA. Blockchain-and-Internal-Control-The-COSO-Perspective-Guidance.pdf

Deloitte (2019). Government jobs of the future. What will government work look like in 2025 and beyond?, USA. DI_Career-coach.pdf (deloitte.com)

Desmet, D., Duncan, E., Scanlan, J., Singer, M. (2015). *Six building blocks for creating a high-performing digital enterprise*. Available via McKinsey & Company. http://www.mckinsey.com/businessfunctions/organization/our-insights/six-building-blocks-for-creating-a-high-performing-digitalenterprise

Drucker, P. (1994). The theory of the business. Harvard Business Review, September–October (pp. 95–104).

Drucker, P. (2008). *Management*, 4th ed. New York: Collins Business. Fayol, H. 1947. General and Industrial Management (Administration, industrielle ET generale). London: Pitman.

European Commission/EACEA/Eurydice. (2020). *The European higher education area in 2020: Bologna Process Implementation Report*. Luxembourg: Publications Office of the European Union. EC-02-20-828-EN-N. ISBN:978-92-9484-356-2. DOI: 10.2797/756192

Fitzgerald, M., Kruschwitz, N., Bonnet, D., Welch, M. (2013). Embracing digital technology—A new strategic imperative. *MIT Sloan management review research report*. Available via Capgemini Consulting. https://sloanreview.mit.edu/projects/embracing-digital-technology/

Garsoux, M. (2013). COBIT 5 ISACA's new framework for IT governance, risk, security and auditing. An overview. http://www.qualified-audit-partners.be/user_files/QECB_GLC_COBIT_5_ISACA_s_new_framework_201303.pdf

Goodhart, C. (1975), Problems of monetary management: The UK experience. In A. Courakis (Ed.), *Inflation, depression and economic policy in the west*. Totowa: Rowman & Littlefield Publishers. ISBN:978-0389201441.

Hirt, M., Willmott, P. (2014). *Strategic principles for competing in the digital age*. McKinsey Quarterly, May 2014. McKinsey & Company. http://www.mckinsey.com/business-functions/strategy-and-corporate-finance/our-insights/strategic-principles-for-competing-in-the-digitalage

IIA. (2020). *Internal audit competency framework*. IIA Publications. https://iiabelgium.org/wp-content/uploads/2020/10/Internal-Audit-Competency-Framework-002.pdf

ISACA-COBIT 5. (2012). A business framework for the governance and management of enterprise IT, an ISACA framework. isaca.org/cobit/Documents/COBIT-5-Introduction.pdf

Kandeh, H., Alsahli, M. (2020). Effect of big data analytics on audit an exploratory qualitative study of data analytics on auditors' skills and competence, perception of professional judgment, audit efficiency and audit quality. UMEA University,

Department of Business Administration Master's Program in Accounting Masters Thesis in Business Administration III, 30 Credits, Spring 2020 Supervisor: Tobias Svanström.

Kane, G. C., Palmer, D., Phillips, A. N., Kiron, D., Buckley, N. (2015) *Strategy, not technology, drives digital transformation—Becoming a digitally mature enterprise.* London, UK: MIT Sloan Management Review and Deloitte University Press. http://www2.deloitte.com/global/en/pages/technology/articles/deloitte-social-business-study.html

Kharabe, A., Lyytinen, K. (2012) *Is implementing ERP like pouring concrete into a company? Impact of enterprise systems on organizational agility.* Cleveland, OH, ICIS: Case Western Reserve University.

Kohnke, O. (2017). It's not just about technology: The people side of digitization. In O. Gerhard & K. Michael (Eds.), *Shaping the digital enterprise trends and use cases in digital innovation and transformation. Preface* (eBook). Switzerland: Springer Nature. ISBN:978-3-319-40967-2.

Kokina, J., Davenport, T. H. (2017) The emergence of artificial intelligence: How automation is changing auditing. *Journal of Emerging Technologies in Accounting: Spring 2017*, 14(1), 115–122.

Kowalkiewicz, M., Safrudin, N., Schulze, B. (2016). The business consequences of a digitally transformed economy. In G. Oswald & M. Kleinemeier (Eds.), *Shaping the digital enterprise: Trends and use cases in digital innovation and transformation.* Heidelberg: Springer.

KPMG. (2017). Enhancing the strategic role of internal audit. https://assets.kpmg/content/dam/kpmg/ch/pdf/enhancing-strategic-value-of-internal-audit-en.pdf

KPMG. (2019). KPMG Internal Audit: in 2020. Considerations for impactful internal audit departments. https://assets.kpmg/content/dam/kpmg/cn/pdf/en/2019/05/top-10-in-2020-considerations-for-impactful-internal-audit-departments.pdf

Lääts, K., Kask, K., Täks, V., Alexandr, T., Tomáš, L., Kleinheyer, B. (2019). White paper on digital transformation of universities' internationalization process. Collaboration in Higher Education for Digital Transformation on European Business. https://www.chedteb.eu/media/attachments/2019/08/25/white_paper_on_digitalisation_universities.pdf

Markovitch, S., Willmott, P. (2014). *Accelerating the digitization of business processes*, May 2014. McKinsey & Company. http://www.mckinsey.com/business-functions/business-technology/our-insights/accelerating-the-digitization-of-business-processes

McAfee, A., Welch, M. (2013). Being digital: Engaging the organization to accelerate digital transformation. *Digit Transform Rev 4*, 37–47.

Naurem, D. (2020). On the job training methods. https://www.economicsdiscussion.net/job-training/on-the-job-training-methods/31626

Nzechukwu, P. O. (2017). *Internal audit practice from A to Z.* Boca Raton, FL: Taylor and Francis. ISBN:978-1-4987-4205-4.

Oswald, G., Kleinemeier, M. (2017). It's not just about technology: The people side of digitization. In O. Gerhard & K. Michael (Eds.), *Shaping the digital enterprise trends and use cases in digital innovation and transformation. Preface* (eBook). Switzerland: Springer Nature. https://doi.org/10.1007/978-3-319-40967-2_3. ISBN:978-3-319-40967-2.

Protiviti. (2020a). Returning internal audit to "Business as Usual" in the new World. https://www.protiviti.com/sites/default/files/united_states/insights/returning-internal-audit-to_business-as-usual-in-new-world-protiviti.pdf

Protiviti. (2020b). Exploring the next generation internal auditing. https://www.protiviti.com/sites/default/files/2020-ia-capabilities-needs-survey-protiviti.pdf
Salijeni, G., Samsonova-Taddei, A., Turley, S. (2019). Big data and changes in audit technology: contemplating a research agenda. *Accounting and Business Research* 49(1), 95–119.
SAP. (2015). SAP white paper: value creation in a digital economy—Adapt or die in a digital world where the consumer is in charge. https://www.ciosummits.com/Digital_Business_Whitepaper_FINAL_external_09_08_15.pdf
Suffield, M. (2020). Auditors of the future – what are the skills needed in a digital age? *Journal of European Court of Auditors* 1(1), 23–27.
Vasarhelyi, M. A., Soohyun, C., Arion, C., Chanyuan (Abigail), Z. (2020). Smart audit: The digital transformation of audit. *Journal of European Court of Auditors* 11, 27–32.
vom Brocke, J., Rosemann, M. (2014) *Handbook on business process management 2: Strategic alignment, governance, people and culture*. Heidelberg: Springer.
vom Brocke, J., Zelt, S., Schmiedel, T. (2016). On the role of context in business process management. *International Journal of Information Management* 36(3), 486–495.
Wang, X. (2010). Performance measurement in universities. Managerial Perspective. Business Administration – Financial Management Faculty of Management and Governance University of Twente. Supervisors: Prof. dr. Nico P. Mol and Dr. Ben Jongbloed.
Weinert, F. E. (1999). Definition and selection of competencies concepts of competence. https://www.semanticscholar.org/paper/Definition-and-Selection-of-Competencies-Concepts-Weinert/8b88efa9dd5e0a4b605aea6e5e3b9ec640beb089?p2df
Westerman, G., Bonnet, D., McAfee, A. (2014). *Leading digital: Turning technology into business transformation*. Boston, MA: Harvard Business Review Press.

Chapter 6

Executing the responsibilities of internal audit in digitally transforming universities

6.1 KEY PERFORMANCE STANDARDS CRITICAL FOR DIGITAL AUDIT

Quality and performance measurements in the higher education sector are among the strategic topics. Accordingly, the existence of continuous auditing and internal control activities with an approach supported by the senior management is of critical importance to achieve sustainable high performance and to provide quality education in universities (Turner 2011; Bozkus Kahyaoglu and Coskun 2020).

It is necessary to determine the standards to be taken as a basis in performance measurement, to examine whether the targeted quality level has been achieved, and to evaluate the quality comprehensively. One of the themes that will emerge in the process of examining the quality of higher education is that what is presented and accepted as the quality depends largely on what the supposed purpose of higher education is. It is possible to explain this briefly with a "fit for purpose" statement (Chambers and Odar 2015). The aims and functions of higher education have changed greatly over time with the effect of technology and innovations.

It is seen that the meaning of "higher education" has changed a lot in time and the boundaries of this concept have become "uncertain" in recent literature. The first thing that comes to mind when we say "higher education" is the university and they are used as synonymous. However, there is considerable flexibility in the content of the word "university" (Baletic et al. 2020).

From this point of view, universities are beyond surviving, rather present successful role models in terms of providing information and social innovation (Benneworth and Cunha 2015). That makes them one of the most successful institutions of the modern age. However, the point to be noted is that when compared with the previous period, what is quality in universities has changed significantly today.

It is important to focus particularly on changes in the way quality and performance criteria are understood and measured in the higher education sector and all universities. Thus, it may be possible to ensure that contemporary standards are correctly understood, and the applicable ones are

DOI: 10.1201/9781003093008-6

prioritized for "quick win". This approach can also help clarify the position of internal auditors concerning the new era with digital audit requirements subject to the changing standards (FINEEC 2017).

Determining who should make the "Quality" decision in the higher education sector is closely linked to the problem of selecting appropriate performance criteria for this assessment. In general, academics' approach to quality may be based on valuing the collective judgments of their colleagues (DeSimone and Rich 2019). However, in such a competitive environment, market analysts may place more value on the monetary dimension. This leads to the question of how to determine performance indicators or objective measures that can be used in evaluating quality in higher education. It should be noted that a performance indicator that can be clearly and precisely defined does not guarantee the meaning of the indicator is uncontested.

It will be necessary to look at other stakeholders who may have legitimate expectations for the quality of higher education and whether they can make an appropriate decision in line with their purpose.

Most comprehensively, prospective students have the most important quality assessment perspective in this context (CohnReznick 2020). However, it should be kept in mind that prospective students have a very different perspective on the criteria they believe are appropriate for the quality and evaluation of higher education. Here, it has the potential to evaluate the effectiveness of the curriculum as the segment that represents the new generation who knows the technological advancements and innovations most closely and wants to implement them, as well as investing in their personal development and careers.

Today, the added value of internal audit in higher education institutions is becoming the focus of attention of all stakeholders due to the increase in digital regulatory requirements in the global competitive environment and the focus on governance, risk, and security issues (IIA 2010; Bozkus Kahyaoglu and Coskun 2020; Türetken et al. 2020).

Considering the traditional IA function, the priority is given to the audits in the field of financial control and compliance, about the protection of assets in general (Allegrini et al. 2006; Dellai et al. 2016). However, with the digital transformation, important changes have occurred in the internal audit priorities of higher education institutions (Erasmus and Coetzee 2018; Bozkus Kahyaoglu and Coskun 2020; Türetken et al. 2020).

According to Mihret and Yismaw (2007), achieving sufficient benefit and added value from the internal audit is closely related to the competence of internal audit as well as how the organization in question is managed. It should be noted that it becomes a necessity to focus on how to add value to the whole organization rather than just focusing on the internal audit function (KPMG Audit Committee Institute 2016). While measuring the performance of internal audit, it is seen that there are quite a lot of studies in the literature regarding its effectiveness and efficiency (Aksoy and Kahyaoglu 2013;

Endaya and Hanefah 2013; Badara and Saidin 2013; Badara and Saidin 2014; IIA 2016; 2017a; Bozkus Kahyaoglu 2017; Lenz et al. 2017; Lenz et al. 2018). In these studies, comparisons are made for different countries and sectors, generally based on IIA standards. Although these studies have important contributions to the literature, there is a need to update and review performance indicators as a result of changing audit and risk universes due to digital transformation (Nurdiono and Gamayuni 2018; Bozkus Kahyaoglu and Coskun 2020; KPMG 2020a).

According to Türetken et al. (2020), improving the quality of internal audits should be considered an important and priority issue. These authors argue that as a key element to demonstrate the digital drivers that internal audit should have, there is a need for "accurate", "realistic", and "simple" key performance indicators to help organizations to measure the effectiveness of their internal audit functions.

It is essential to establish performance standards and related measurement criteria based on the business environment (D'Onza et al. 2015) to measure the extent of internal audit's achievement of its strategic goals and especially the effectiveness of digital audits. It is not sufficient to simply specify the potential effectiveness factors of IAs for the higher education sector. It is also necessary to regularly assess their actual impact on the organization through well-defined indicators. The professional standards of the Institute of Internal Auditors (IIA 2020) provide limited assistance in this regard. Therefore, to contribute to the literature, we need to broaden our understanding of the state of the research on the effectiveness of IAs. Thus, an important step will be taken to examine the indicators or performance metrics used to operationalize/quantify the effectiveness of IAs (Jiang et al. 2016). Especially in this context, it will be easier to understand the factors affecting IA effectiveness based on a comprehensive and up-to-date synthesis relevant for applying in a digital environment.

In the digital age, main activities and processes have been transferred to automated systems in most higher education institutions. In addition, as control activities are also embedded in the system, data governance and control culture become more important. Beyond that, some level of professional skepticism must remain to protect systems from cyber-attacks and determine when the system might go wrong. In this context, the internal audit function plays a strategic role in establishing a data-driven control culture and maintaining a skeptical mindset by implementing a digital auditing approach (Chung 2020).

6.1.1 Requirements for the effectiveness of digital audit

Türetken et al. (2020) argue that even though the major factors affecting IA effectiveness and efficiency are observed in most of the available studies in the literature, these studies do not provide actual indicators or performance metrics. In this context, the performance indicators suggested by these

Requirements for Effectiveness

- R 1 Competence of the IA Department
- R 2 Size of the IA Department
- R 3 Organizational Setting
- R 4 Scope Limitation
- R 5 Compliance with Applicable Standards
- R 6 Management Training Ground
- R 7 Auditee Attributes
- R 8 IA Independence
- R 9 IA Objectivity
- R 10 Conduct Risk Consulting
- R 11 Outsourcing IA
- R 12 Quality of Audit Work
- R 13 Chief Audit Executive's Leadership Style

Targets for Effectiveness

- T 1 Management Support for IA
- T 2 Interaction Between Internal and External Audit
- T 3 Cooperation with Audit Committee
- T 4 Information and Communication
- T 5 Existence of a Follow-up Process
- T 6 Supportive Control Environment
- T 7 Cultural Dimensions

Figure 6.1 Metrics for the requirements and targets of internal audit effectiveness in the digital era.

Source: Türetken et al. (2020).

authors are presented in Figure 6.1, including the requirements (supply) and targets (demand) side of IA effectiveness.

- R 1: "*Competence of the IA Department*"
  Today, the requirements of audit professions are changing based on technological advancements. The main reason for this change is that the tools the auditors use for the audit dramatically change. The changes in the audit tools also affect the algorithms of the internal audit depending on the development of technology. From this point of view, it becomes a requirement for internal auditors to have the ability to audit business processes related to the use of technology, and consequently, to engage in continuous learning activities for innovations. In

this sense, technological competence, which seems to be a different field, is a necessity that internal auditors should have. Especially since universities are places where technology is utilized widespread and turned into best practices, it is necessary to have a high level of knowledge in this regard as well. It should be noted that the HE Institutions are institutions where decisions that will directly affect the lives and careers of individuals are made and information related to this is produced. Considering the decisions to be made in this respect may have irrevocable and irreversible effects, the HE institutions have such institutional structures where the audit is the most crucial need. These factors can be seen as the source of financial risks that may arise in general and most importantly as the source of risks that may lead to damage to the reputation of the HE institution (Abraham and Walker 2017). Therefore, internal audit should be seen as the most important tool, especially for managing reputational risk.

- R 2: *"Size of IA Department"*
  Audit activities are evaluated depending on the magnitude of the losses that will occur when there is no financial advantage for the HE institutions. Therefore, the rule of thumb in establishing an optimal internal audit department will be determined according to the situation where the marginal cost incurred as a result of the expenditure incurred for the department is equal to the marginal return, which is an advantage for the organization (PwC 2014), and hence this corresponds to the losses prevented by the IA department.

  From this point of view, since the risk of loss arises depending on the size of the HE institutions, internal audit departments will also grow depending on this scale. As the size of the HE institution increase, the relatively faster increase in the size of the internal audit department means that "negative internal economies of scale" (Liang-Cheng 2016) is generated for the HE institution. For this reason, as the size of the HE institution grows, the number of units to be audited and the risk of losses should be evaluated according to the benefit obtained from the audit fieldwork activity.

  Since internal audit and internal control processes in the services versus production sector operates differently, as they are dependent on the diverse business processes, they should be evaluated differently.

  It is a fact that the HE institutions have a complex structure (Ntim et al. 2017). In this respect, internal audit aims to reflect on managing financial risk and ensuring efficiency in resource allocation in different faculties, departments, and institutes. The complex structure in HE institutions necessitates that the personnel who will audit this also have a complex ability and competence (Li and Chen (2012). Therefore, it is necessary to audit larger systems with fewer auditors. While technology provides this opportunity, the use of technology also reveals "positive internal economies of scale" for the internal audit.

For HE institutions, technology is the most important element that creates both positive internal economies of scale and negative internal economies of scale. The internal audit is a mechanism especially for the management of internal economies of scale. However, depending on the growth of the HE institutional structure, the obligation for the IA department to grow also leads to a "negative internal economy of scale" (Ross 2019).

Universities have the advantage of operating in various fields as a "scope of economies" (Agasisti 2016). In particular, the economic value and social benefit of the information they produce are high. However, these scopes of economies are also the main reason for the necessity of continuous monitoring and auditing as a critical process. Therefore, as the scope of economies in the universities is directly related to the growth rate of the universities, it becomes a necessity to construct the internal audit department accordingly.

In general, the methods, approaches, and techniques used by internal auditors have emerged as a tool to manage the resources of the HE institutions aiming for minimizing the effects of the economic crises. For this reason, the audit understanding of the HE institutions has emerged as a structure for auditing the accounting and financial processes (Ntim et al. 2017). However, developments, especially technological developments, have made internal audit the most important tool of risk management.

In today's university structure, it is seen that internal audit processes are generally structured toward more effective use of resources which is far beyond accounting and finance. The technological developments and the momentum that it aroused will lead to the differentiation of the tools, techniques, and methods of internal audit (Peterson and Carl 2014). This new period can be defined as the period after COVID-19.

- R 3: "*Organizational Setting*"
  The structure of the HE institutions is determined by the rules and regulations of domestic law. However, internal audit focuses more on processes rather than structure. As processes come to the force more, process-based audits are more important than a traditional audit approach (Baker Tilly 2019). In this context, ethical codes also gain more importance. Processes are moved from the dimension of physical space to the virtual arena with the effect of digitalization. This situation causes the control mechanism to expand and the workload to increase for internal auditors.
- R 4: "*Scope Limitations*"
  According to Erasmus and Coetzee (2018), one of the most important aspects of the effectiveness of the internal audit is that there is no scope limitation. Thus, there are no restrictions on IA activities. It means that internal auditors can investigate any process, system, and document within the organization and communicate with all relevant

stakeholders within the framework of risk-based audit plans. As a result of an accurate scope-based internal audit activity, recommendations for potential process improvements can be offered, or nonconformances can be detected early to deter potential fraud cases in the HE institution.

Auditors can't audit areas where they are not competent. Therefore, scope restrictions may be encountered. In this case, although outsourcing is considered as a solution, it should be taken into account that it will not be sustainable (KPMG 2016).

- R 5: "*Compliance with Applicable Standards*"
  The higher education sector is generally positioned among regulated markets. In this context, it is expected that the internal audit team will carry out fieldworks that will provide sufficient assurance for the full, complete, and timely realization of all legal requirements that higher education institutions must comply with. As a result of digitalization, legislation regarding the ownership and protection of data is among the sensitive and priority issues (PwC 2015, 2018; Betti and Sarens 2020). Therefore, internal auditors should closely monitor the legal needs and review the amendments and closely watch the level of compliance with the corporate legislation.
  For this, the performance of the internal audit department and all its activities following the internal audit standards should be considered as a prerequisite. The IIA standards do not provide complete guidance on the business model of higher education institutions at this stage, but they provide very comprehensive information and advice and set examples of best practices (Dejnaronk et al. 2016).
- R 6: "*Management Training Ground*"
  The internal audit team can raise awareness to increase the effectiveness of internal audit activities within the organization through training and thus contribute to cultural development. However, some researchers point out (Yee et al. 2008; Dellai et al. 2016) that there may be some drawbacks in its implementation. Internal auditors can provide great benefit in training the university management, provided that their corporate objectivity is not compromised.
- R 7: "*Auditee Attributes*"
  The most important feature of university officials should be to behave ethically in their work and activities. When evaluated from this perspective, internal auditors should generally attach importance to ethical audits in their risk-based audit plans. Considering that HE institutions are an area where human-to-human relations are intense, ethical controls become the most important tool to prevent any attributes in the business processes which may intend to provide unjust benefits to individuals (KPMG 2020b). Although digitalization is a different dimension of ethical controls, it increases the necessity of internal auditors to watch the attributes of auditees more closely.

Continuous testing and monitoring of controls is an essential issue to provide the necessary assurance in higher education institutions. For this, interdisciplinary teams and audit experts must be employed in the internal audit department. Thus, internal auditors must make a plan to test controls and to continuously audit the processes of business process owners. In particular, it requires continuous monitoring of the appropriateness of the technical personnel's skill sets to create a capacity to audit technology-supported control systems (Bozkus Kahyaoglu et al. 2020; Vasarhelyi et al. 2020).

Technological change is shifting the business processes in the universities within this framework. People demand products and services from the universities that will determine and improve their future standard of living (Bailey 2016). In this respect, depending on this new situation, universities will focus on their internal structure to define their strategic directions. Thus, they may decide to include the whole of the university or they may include some of their units or processes to provide the service/teaching style they will provide.

The structuring of the audit departments should be established accordingly to analyze whether the existing decisions are effective both at the university level and in the context of the service provided. It should be noted that complex education programs that will turn toward a more interdisciplinary field will further increase the need mentioned above (Bozkus Kahyaoglu and Coskun 2020; KPMG 2020b).

- R 8: "*IA Independence*"
  It is a fact that management must deal with a fast-changing business environment. As trusted advisors, internal auditors should initiate a mindset that is more oriented toward advisory roles to guide the HE institutions regarding strategic changes in the digital era.

  Internal audit departments should present how they add value to auditees to get a better understanding of the digitalization process of the business and to ensure that they contribute to organizations. However, this issue represents a challenge for internal auditors regarding maintaining their independence. Betti and Sarens (2020) draw attention to the difficulties internal auditors face regarding their autonomy within the organization due to their consulting activities because of digital transformation. These authors state that the independence of internal auditors is a very decisive issue. Therefore, it reveals that internal auditors should consider balancing advisory activities in their risk-based audit plans (Kane et al. 2016).
- R 9: "*IA Objectivity*"
  Considering the analysis of data and data governance framework of higher education institutions today, it should be kept in mind that data analytics can have "disruptive effects" on the relationship between auditors and auditees. According to the joint work of PwC, ACCA, and INSEAD (2019), the most important way to contain these disruptive impacts is to define clear roles and responsibilities for data source

and access, data information, and data quality. In this context, a collaborative approach between process owners and internal auditors should be considered. There is a growing need for an objective approach from auditors to any business process or transaction that requires judgment. This situation becomes more important in the provision of consultancy services in the digital transformation process.

In the digital transformation process, it is important to determine the priorities in the automation of the relevant business units and processes in terms of achieving financial goals and resource management. In this context, a research report by Deloitte (2018c) argues that internal auditors are in a key position in determining the priorities of the organizations in the transition to automation and that an objective perspective of internal audit is essential for a value-added approach.

- R 10: "*Conduct Risk Consulting*"
  The internal audit approach for assuring regarding the past events and transactions has been replaced by providing assurance based on a risk-based approach in the digital era. Increasing the digital competencies of internal audit (Kamusoko and Jingura 2019) has an important role in realizing this approach.

  Internal audit must develop a perspective beyond the traditional risk management structure defined within the "three lines of defense" (Deloitte 2020b) considering the needs and expectations in the digital era. With this approach, it is only possible for the higher education sector to detect new risks early in the rapidly increasing risk portfolio to execute a successful integrated risk management strategy (Lenz and Hahn 2015). Hence, internal auditors need to use technological developments effectively, especially in risk analysis, measurement, and evaluation processes within the HE institutions. For this, they inevitably show continuous improvement to maintain their role as trusted advisors.
- R 11: "*Outsourcing IA*"
  In the performance of internal audit activities, it appears as a preference in the form of outsourcing or a combination of both, as well as using internal resources. Dellai et al. (2016) reveal the various advantages and disadvantages of outsourcing of internal audit functions. In this respect, it is stated that internal audit activities performed with the use of internal resources are easier to acquire skills and knowledge in the audited area. On the other hand, there is an acceptance that outsourced internal audit services increase audit objectivity.

  The cost of setting up an internal audit department within the organization is generally considered to be higher than outsourcing, mainly considering the costs incurred in recruiting and training the audit team. On the other hand, in the internal audit activity provided by external sources, the possibility of the auditors lacking critical information about corporate strategy, the business model, and the corporate culture is considered as a source of risk (D'Onza et al. 2015; Türetken et al. 2020).

It is a fact that the existence of an experienced audit team with competence in the digital transformation process is important in the good management of the process and guaranteeing objective decision-making. Therefore, the level of advantages and disadvantages should be discussed beforehand to achieve a balanced preference after evaluating the specific conditions of the HE institution. It should be noted that the key determinants of this decision are the audit committee and the board of directors (Deloitte 2020a).

- R 12: "*Quality of Audit Work*"

  It is a fact that internal auditing is defined as a "knowledge-based activity". For this reason, internal auditors should be proficient in identifying and analyzing the data to obtain a depth of knowledge about processes and activities in the HE institution. In connection to this, the quality of internal auditing improves directly with the degree to which such a knowledge is shared and used by the stakeholders. On the other hand, Pitt (2014) argues that knowledge is valuable when it is managed conveniently.

  Chief audit executives should arrange a "quality assurance and improvement program (QAIP)" for the internal audit department, regardless of the size (Sarens et al. 2012). The concept of quality in internal audits requires an approach beyond compliance with the standards. Rather than this, the key determinant of quality is based on stakeholder expectations. Hence, it should be performed by the internal audit department as a way to contribute with a value-added approach for the HE institution. The concept of value can be different for each organization. Therefore, chief audit executives should determine what will be perceived as both value and quality for their organization and assign strategic direction accordingly.

  It is necessary to group each of the key inputs which can assist internal auditors to manage and measure their quality. Such a grouping can be done by defining inputs for the strategy and budget, staffing, and professional practices relevant for the HE institution.

  Chief audit executives should undertake regular monitoring activities for the inputs are given in Figure 6.2 to continuously improve the internal audit function and demonstrate the performance of internal audit transparently.

- R 13: "*Chief Audit Executive's Leadership Style*"

  One of the factors determining the value of internal auditing is whether the CAE has leadership qualities (D'Onza et al. 2015). In the literature, leadership is seen as the most important skill that CAE should have (Bednarek 2018). Among the studies frequently encountered in the literature, the importance of the quality of leadership demonstrated by the CAE in determining and executing the IAE and the impact of CAE's leadership style on IAE effectiveness are emphasized.

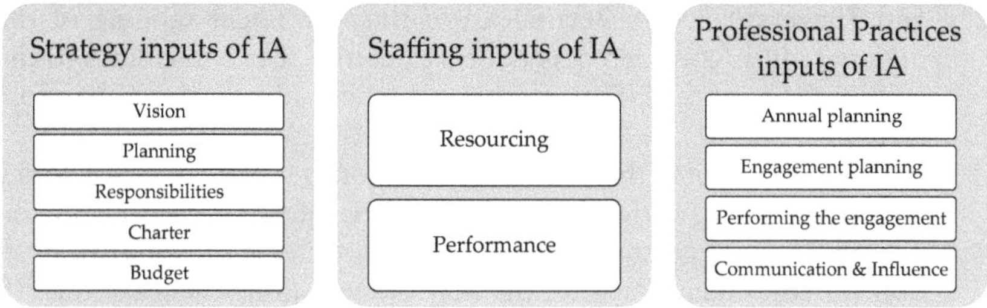

Figure 6.2 Inputs for internal audit quality measurement.

Source: Pitt (2014).

6.1.2 Targets for the effectiveness of digital audit

In this section, details of the target factors are presented which have an impact on internal audit's effectiveness. The literature discusses the basic characteristics that an effective IA to achieve its goals should have in seven key categories. These are briefly explained below. However, Erasmus and Coetzee (2018) detect a widening gap between the current stakeholders and the expectations of internal audit at the level of target realization in the digital era.

- T 1: "*Management Support for IA*"

  Within the scope of IIA standards, internal auditors act with independence, impartiality, and objectivity while performing their duties. However, some works argue that they have a limited capacity to fulfill these duties despite IIA standards (Alzeban and Gwilliam 2014; Pitt 2014; Alshbiel 2017; Bednarek 2018). What is meant here is to demonstrate the importance of management support. Management support is critical in facilitating the fulfillment of their duties by internal auditors and their firm stance in implementing the recommendations contained in audit reports (Halimah et al. 2009; Baheri et al. 2017).

  The historical background and experiences on the risks and the risk management approach cause a change in the understanding of auditing (McKinsey 2017a). Contemporary risk management has emerged from the experience of crises and challenges. In this context, the development of technology has accelerated this process.

  This technological change, at the same time, brought along integration of risk management perspective. From this point of view, the understanding of internal audit has become a valuable tool for risk management. This new approach is aimed at preventing events before they happen and has gained rapid acceptance as best practice. In addition, the acceptance and support of the new audit approach, i.e., "digital audit", at the management level has brought more flexibility in process management of the HE institutions (Struthers-Kennedy and Fitzgerald 2019). Beyond all of this, the understanding of the digital audit is considered as the best practice that contributes to the sustainability of the HE institution, with high added value (Chandola 2015).

It should be noted that the above-mentioned factor has revealed the advisory function of the digital audit approach. This advisory function also reveals a trend toward having a responsible representative position for risk management in the new organization scheme of the HE institution. That is because of the necessity of protecting extensive resources, especially against third parties (EY 2018), which has ensured the institutionalization of this new understanding and has been seen as an important experience. This new experience in the HE institutions has also turned into general practice with the audit fieldwork and digital adaptations process.

- T 2: "*Interaction between Internal and External Audit*"

  It is a fact that there is an increasing interaction between internal and external audit activities compared to the previous period. It is possible to express these interaction areas as joint audit planning, information sharing, and report exchange, etc. There are findings in the literature regarding the positive effect of this interaction in terms of both internal audit and external audit (Alzeban and Gwilliam 2014; IIA 2017b; Chartered Institute of Internal Auditors 2020).

  It should be noted that although internal and external auditing techniques are almost identical, their "intended outcomes" are different. According to the IIA (2017b), internal audit's mission is "to enhance and protect organizational value by providing risk-based and objective assurance, advice, and insight". In this respect, the internal auditor attributes priority to determine whether an organization's operational practices are encouraging the business to achieve its goals, by identifying and managing the key risks.

  Given the structure of the higher education sector, audit committees need more operational knowledge. This operational knowledge can be obtained through the internal audit mechanism rather than the external audit. Internal audit plays a functional role, especially in human behavior, ethical rules, and access to sensitive data in various core business processes and sub-processes. However, the external audit also has a significant complementary role (IIA 2017b). In this context, the requirements of the "three lines of defense" are fulfilled (Oliver Wyman 2020; RSM 2020).

  In the study conducted by Deloitte (2020a), it is revealed that there has been a significant change in the risk universe, and it has turned into a very complex structure. The traditional three lines of defense remain weak in the early detection of risks at this digital risk universe. Standards were updated by the IIA to eliminate this deficiency in July 2020.

- T 3: "*Cooperation with Audit Committee*"
When evaluated in general, the internal audit department and the audit committee are defined as two different control authorities. The internal audit department performs audit and consultancy activities within a higher education institution. On the other hand, an audit committee is formed within the Board of Directors for monitoring and evaluation based on the information taken from internal audits and other stakeholders (Arena and Azzone 2009). Therefore, the cooperation between the internal audit and the audit committee has significant effects on the effectiveness of internal audit activities through the sharing of information and reporting. In this respect, cooperation and solidarity between both control authorities can make an important contribution to the enterprise-wide risk management process (Alshbiel 2017; Bednarek 2018).
- T 4: "*Information and Communication*"
Information and communication include achieving internal audit objectives by the COSO conceptual framework. Accordingly, it refers to the sharing of information and opinions obtained within the scope of the audit in an appropriate and timely manner by the internal audit department (Lenz et al. 2017). In addition, communications on internal audit should include predetermined strategic goals, audit findings, results, recommendations, and action plans (IIA 2017a; IIA 2017b).
  It considers both verbal and written communication of the internal audit department with relevant stakeholders in the university including the audit committee, the board of directors, Senate, and Deans). It should be noted that the audit report is taken as a formal communication channel for internal audit and as part of communication with the management level (Dejnaronk et al. 2016).
  Due to the rapid increase of digitalization, there may be a need for an update in the auditors' perspective regarding information and communication. However, the change and development of technology do not mean that internal auditors will do other things. Therefore, they are expected to complete their work in a better, faster, and higher quality and present the audit findings to the management with a value-added approach. In this process, a change in the form of information and communication may occur depending on the level of institutional maturity (Beni 2019). Internal auditors should pay special attention to the information and communication, whatever the circumstances to achieve their strategic goals.
- T 5: "*Existence of a Follow-up Process*"
From the standpoint of internal audit, the follow-up process is not limited to an assessment that the relevant recommendations have not been chased (IIA 2017a). It requires an approach beyond that. Particularly, it is important to consider whether the auditee has adequately addressed the issues raised through the audit. In this context,

it should be noted as a key difference in approach, whether it is a superficial point of view in solving the problems or whether a permanent solution by internalizing it. At the same time, it is not enough to admit the existence of the problem (Türetken et al. 2020). It is necessary to monitor whether the underlying situation is resolved under strategic objectives after a reasonable period (Oussii and Taktak 2018).

- T 6: "*Supportive Control Environment*"
  With the digitalization in the higher education institution, a modernization that allows real-time monitoring is emerging in the first and second line of defense, i.e., internal control environment. Although the internal audit department cannot fully rely on the work performed by the first and second lines of defense in the HE institution, they can use it for CA/CM activities (Vasarhelyi et al. 2020). Thus, it can contribute to the early detection of risks and focusing on more critical risks. This issue becomes very important in the digital transformation process in the HE sectors to create a supportive control environment.

  However, it should be noted that there may be some cases where not every department has completed the digital transformation at the same level. That can lead to different digital maturity levels in different processes or different units within the same university (Deloitte 2018b, 2020b). It would be appropriate for the internal audit department to analyze digital maturity levels first and provide assurance or consultancy services accordingly in such a situation.
- T 7: "*Cultural Dimensions*"
  In the digitalization process, there is a need for the transformation of human resources and technological transformation (McKinsey 2017b). The structure of corporate culture is important for the effective realization of this transformation process. Without considering the cultural dimension, the process of change and transformation is almost impossible to be successful. Therefore, a strategic role is assigned to ethical codes and corporate values to develop the constructive and collaborative structure of the corporate culture in the digital era, considering the HE sectors. In this context, internal audits can contribute to shaping the necessary corporate culture by offering constructive recommendations.

  A strong corporate culture in line with a business strategy set in competitive conditions in the higher education sector is highly effective in accelerating the performance of the higher education institution. It is also an important factor promoting the ability to outpace competitors, especially in this digital era (Alzeban 2015).

  The most prominent features in the corporate culture of the digital era are determined as high speed, cooperation, and delegation. In this context, internal auditors undertake an important task in developing adaptation and inter-process cooperation and identifying risks early in the transition to a new cultural structure. For this, the transformation of internal audit infrastructure and the application of CA/CM audit tools and techniques can be expressed as a prerequisite (Vasarhelyi et al. 2020).

6.2 BALANCING FUTURE RISK VISION OF MANAGEMENT AND INTERNAL AUDIT

The current audit approach is established after the financial crises which focus on defending against risks especially arising from the errors of institutions. However, in practice, this defensive approach in audit understanding leads to a loss of "flexibility" in the organizations. At the same time, it restricts the tendencies toward the expansion and innovation process by having a role in risk aversion attributes at the institutional level (Jiang et al. 2016; KPMG 2016, 2020a).

In general, a differentiation in roles has emerged in the understanding of internal audit and the functional separation caused by this differentiation has become an important problem, especially related to the added value expected from internal audit. In particular, the opportunity created by advanced technology requires a synchronous management style in which the duties or roles of those involved in all business processes operate simultaneously (Zipper 2016).

Digitalization brings out transparency and consequently accountability as a necessary outcome in the management approach. In this sense, it is necessary that everyone, from the smallest unit of institutions to the top management, should adopt an objective and clear accountability as a culture (Fonseca et al. 2020). This process will be one of the most fundamental approaches to the new risk understanding in the HE sectors. However, especially in the management approach of HE institutions, it is necessary to develop behavioral criteria in a way that can achieve the goals of stakeholders and to attach importance to practices for the adoption of ethical rules.

To achieve a corporate purpose, the understanding of audit should assume its role that it is effective in the universities' core business areas and develop this role with its contribution to risk management (Christopher 2012). In this respect, modern internal audit's support for assurance and consulting should be extended to include governance and risk management (IIA 2020).

At this point, whether the auditors come from the processes of the institutions or not is a challenge. For the internal audit approach to be objective, they should not have any managerial role and be independent of operational responsibilities (Lindberg 2013; KPMG 2020a). For all of these to happen, it is a necessity for the owners of the responsibilities related to audit and audit to be aligned with their interests. In this respect, the vision of the new internal audit should be based on an approach to create value and protect it (Fonseca et al. 2020).

It is a fact that technological developments reduce human impact and effectiveness in business processes. Therefore, the control of processes becomes the most important audit area, especially in areas where technology is used extensively. At this point, a key role of a data-based audit approach in business processes comes to the fore. In this case, determining which indicators reflect the whole process in business processes will determine the effectiveness of the audit in terms of monitoring, surveillance, and risk management.

Considering that each of these indicators is also a risk indicator, the data production process becomes the first stage of risk management. For this reason, data acquisition, analysis, and the fact that an audit process creates information about what needs to be done as a source of information transform the auditor into an expert who is competent to perform data analysis.

Especially in the universities, the structure of auditing is aimed at auditing the core business processes. In this context, this general operational framework necessitates the expansion and redefinition of the basic functions of internal auditing in the HE sectors. In this respect, the new audit approach is rapidly transforming into a consultancy and risk management structure. In other words, the whole audit processes turn into a protection activity against reputation risk, especially for universities.

Since the reputation risk is a result of many components, it is necessary to establish the analysis and indicators of the determination of these components and to create continuous reporting information by making use of them. In this process, the output of the data coming from within each unit as if a neural network, as an interrelated network has turned into a structure that creates the "reputation" of the university (PwC 2006; Origami Risk 2020). The detection of hidden layers and latent variables within this network structure becomes the basic rule of effective risk management. In this sense, the internal audit department represents the structure that forms an audit data universe regarding the university business processes.

Considering that the specified hidden layers and latent variables are stochastic, it is also true that this structure itself has an interaction process dependent on conditionalities. In this respect, it is necessary to be sensitive in data collection and analysis, considering that a small impact is a bigger risk factor than itself (Deloitte 2017).

Because of this structure, a result of the complex integrity of the system, it becomes important to establish an audit process based on the "entropy" structure of the data. The entropy structure can be expressed as a process of capturing information to prevent the risk of loss to be caused by any event or phenomenon, rather than the result being known when an event occurs. The ultimate purpose of internal audit in today's conditions is to focus on this information and capture it as soon as possible (Martel-Escobar et al. 2018).

6.3 PERFORMING CONSULTING TYPE ENGAGEMENTS: BENEFITS AND CHALLENGES

With the effect of digital technologies, the structure, system, and processes are transformed in the higher education sector and all universities (Bozkus Kahyaoglu and Coskun 2020). This situation also closely concerns and affects the internal control procedures, general control environment, risk management, and audit practice in the corporate structure (PwC, ACCA, and INSEAD 2019). It should be noted that the corporate maturity level of the HE institution and the maturity level of the internal audit department will determine what has been done and what needs to be done by the internal audit.

Although digital transformation is an important agenda item in the HE sector, it is impossible to think that the same level of transformation takes place in all universities. High efforts are made by finance and core operations processes to automate controls and increase precision in the HE institutions. Some higher education institutions are starting to use sensors to monitor the security of their campuses, all their facilities, and even all their operations. In various universities and colleges, financial areas are prioritized and hence, distributed ledgers are introduced to keep track of all financial transactions of the organization from end to end. In some universities, Robotic Process Automation (RPA) is used. In addition, there are higher education institutions that continuously monitor and visualize corporate risks in real-time through Artificial Intelligence (AI). However, with the rapid development of digitalization, there are new risks, especially in the field of cyber security and data privacy (Deloitte 2018c; Baker Tilly 2019; COSO 2020; Deloitte 2020a). To control these new risks, internal audits should begin to provide assurance or consultancy services, taking into account the maturity level of the institutional structure, systems, and processes.

According to COSO (2020), internal control standards are still operable and relevant in the digital business environment. It is a fact that advanced technology such as blockchain, AI, and RPA improve internal controls and makes them even more, economic, effective, efficient, and extensive.

In the digitalized higher education sector, it places more control on the automated systems in order to increase the efficiency of the complex business processes of universities and gives importance to data governance. In this context, the "control culture" of higher education institutions will become more important in the new digitalized age. Beyond this, the importance of internal audits will also increase. In other words, a certain level of "professional skepticism" is needed to challenge the university's comprehensive systems that consistently generate sensitive data and to identify when the system might go wrong (Ciolek 2017). The strategic function of internal audit plays an important role in both establishing a data-driven control culture and maintaining a skeptical mindset.

As an important effect of digitalization, one of the important responsibilities of the internal audit department is the continuous testing and monitoring of internal controls. While performing this task, working with an interdisciplinary approach becomes a fundamental requirement. Accordingly, while audit professionals in the internal audit team ensure the testing of internal controls, they require the formation of a skill set that includes a wide range of technical competence.

While the application areas of new technologies are spreading rapidly in all sectors of the world, it is a fundamental requirement that the higher education sector always is "one step ahead of other sectors" (Cangemi 2015) to carry this vision further. In general, it is seen that the top management focuses more on the expectation of increasing growth while demanding technological innovation. However, to obtain satisfactory benefit with a holistic approach, it is an important step that core business processes and the risk and control functions benefit from these technological innovations.

The literature shows that the most fundamental challenge higher education institutions must overcome is not the technology itself. In fact, the source of this difficulty stems from the failure to fully adopt the technology within the institution.

Considering the HR structure of HE institutions, it is determined that some people show a tendency to resist the change because of the fear of change. Others may not be able to effectively use the new systems and tools that the HE institution has invested to achieve its strategic goals due to a lack of skills or training. In such a dynamic business environment, a well-thought-out change program is needed that is supported and guided by the top management (Allui and Jolly 2016). At this stage, internal audit must assume a consulting role, transforming digital control functions and preparing them for the future.

In the joint study conducted by PwC, ACCA, and INSEAD (2019), in the examination of the five components of COSO's 17 principles, it is stated that an institution can have a more effective internal control environment if it has a properly used technological infrastructure in line with its strategic objectives. The key principles in the COSO conceptual framework do not change with the application of technology. It should be noted that technological developments do not mean that they do not change the internal control environment. In other words, with the help of technology, the availability of big data can be attained in virtual platforms and networks. In this context, internal control infrastructure can be designed and implemented specifically for the organization following automated procedures. Furthermore, the same principles of COSO are relevant and can be applied appropriately in all cases, i.e., either manually or fully automated (COSO 2020).

6.4 INTERNAL AUDIT'S ROLE AS TRUSTED DATA ANALYZERS

Given the dynamic nature of the higher education sector, the structure of the data is similarly dynamic and complex. The general feature of this data is that it includes data beyond the organization's books and records, which are called external data. These data are generally obtained through "big data", "IoT", "social media", and "traditional applications" and obtaining a wide variety of data from sources such as other data sources that are too large or complex to analyze. This situation brings along more monitoring and detailed analysis of external data as a necessity for internal auditors. For this, there is a greater need for advanced audit data analytics (Newman 2017).

In general, digital standards have started to change and differentiate in the control of the digital environment. Within this framework, the processes ranging from information and educational materials, communication on the network, processing of courses, and evaluation of results, especially in distance education provided by universities, emerge as a new audit universe. It is a fact that traditional audit tools and techniques are insufficient for determining and detecting risks considering this new audit universe. Therefore, it is essential to apply new and advanced data analysis and audit techniques which are summarized in Figure 6.3. Due to this new audit approach based on a comprehensive analysis of data, internal auditors are now becoming "trusted data analysts" (Bozkus Kahyaoglu and Coskun 2020) beyond their trusted advisory role.

In general, information about "text mining and NLP" and "speech, image, and video analytics" should be included in the reporting of audit findings and recommendations in terms of corporate risk management. A "geospatial analysis" (Wise 2018) process should be produced with approaches to

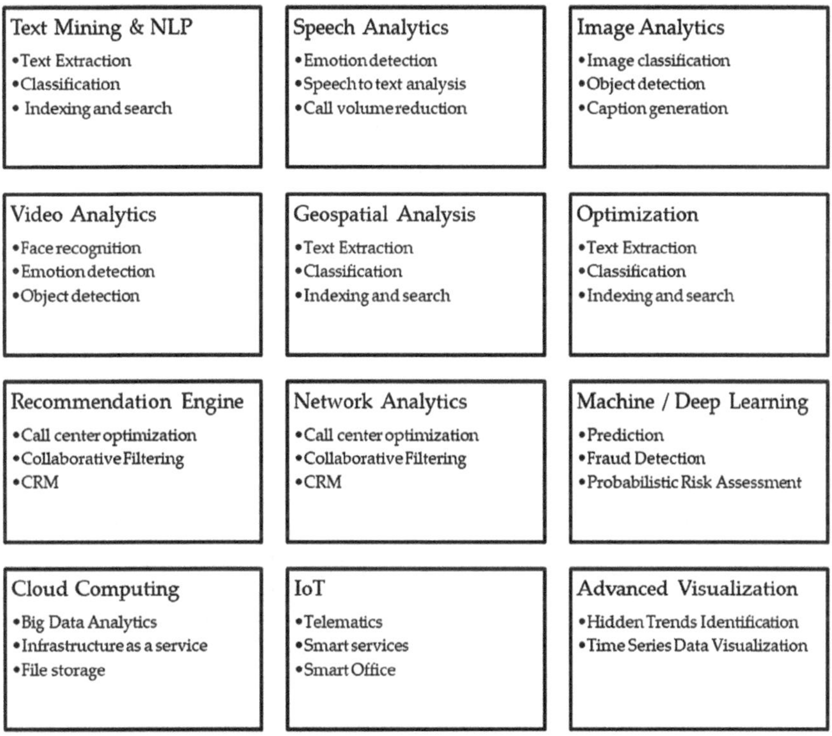

Figure 6.3 Advanced data analysis and audit techniques.

Source: Master and Bansal (2019).

examine the existing digital space as a new geographical area. A "network analysis" will be needed within the structure formed by the links in the virtual platforms of the HE institution. In this context, "artificial intelligence, machine learning, and deep learning" methods and algorithms come to the fore as a tool (PwC, ACCA, and INSEAD 2019).

This process, which has gained importance in terms of reporting, also reveals the knowledge and digital skills that internal auditors should have to achieve the best practice. When evaluated from this perspective, it is seen that internal audit has evolved toward a focus and concentration on the areas where there are more corporate risks, such as cybersecurity risk, emerging technology risks, and HR competency risks (Grant Thornton 2017). This process requires a transformation in internal audit structures within the HE sectors. The data-based process in the HE institutions requires governance over data. This requires the internal audit team to be intelligent. Thus, internal auditors convey the strategic actions to be taken against situations that are likely to happen in the audit reports on a scenario basis. That will strengthen the assurance and consulting aspects of internal audit.

This will apply to all university stakeholders. The most crucial point at this point is that digital information can be used at any time. It is also the real-time generation of this information. Therefore, the audit should be structured to be continuous and real-time depending on this information structure.

Universities are the basic institutions where the information in question constantly flows and leads to the formation of big data. Considering the demographic structure of these institutions, the students who are the buyers of this process are also the participants of this process. Their dynamism and accumulation may emerge as a risk factor in these processes, within the possibilities provided by digital opportunities. In this respect, in the effective control of this dynamic process, the knowledge of those who know all control systems and constantly audit them should be ahead (Green Book 2017). In this regard, applications prepared with new technologies, blockchain, and artificial intelligence tools will be helpful and facilitating and will lead to the formation of a new audit process and workload (Master and Bansal 2019).

The fact that the information and business processes in the digital environment form a chain and even the slightest break in this chain will occur as a "collapse", changes the control and audit work and the concept of risk. In this sense, the risk assessment and control processes, the establishment of information and communication, and the function of surveillance should be performed as processes that do not differ from each other as a whole.

From this point of view, instead of a "defensive" understanding of risk, it is necessary to create a controlled space that will completely manage risks and prevent their occurrence. The fact that this space is a network consisting of the whole relationship with the invisible and much information will require the detection and analysis of risk in an invisible world in terms of auditing. The most important feature of this world is that information is a large stock of data. This data stock can create an environment where simulations can be made in terms of future analysis.

The most important issue here is to capture the risks that may cause a change in the information under the assumption that the hardware structure does not change within the audited structure. Against the risks that may cause changes in the structures and cause irreversible damage to the systems in general, the audit should be evaluated within the scope of the "security strategy" and defined as a risk. Therefore, in the digital world, the risks arising from the infrastructure and the risks arising from the superstructure are integrated (Kokino and Davenport 2017; Deloitte 2018a, 2020a).

Cooperation with the external public stakeholders against shocks that will affect this structure from the outside will be preventive. However, the information to be given to external stakeholders against these shocks will be formed by the control mechanism. In the digital world, universities will be institutions where the understanding of supervision differs within the framework of a structure that integrates the audit field within the framework of the above-mentioned issues.

In this context, the internal audit system will not only focus on the efficiency and productivity of financial resources but will also be structured to identify those that may cause losses in these resources and liquidity risk, or even cause irreversible corporate effects (Grant Thornton 2017). Therefore, the formation of auditors from those included in the said integrated business processes, in general, may shorten the prevention, intervention, and capture processes. The main criterion in the new internal audit approach will be "time" for the universities. The essential point here is that the basic savings of universities consist of digital assets. This issue will bring about the storage, and control problems of these digital assets.

It is a fact that the responsible and ethical use of rapidly developing technology is a fundamental requirement for both process owners and internal auditors. Especially, this requirement becomes more decisive with artificial intelligence applications. On the one hand, although artificial intelligence provides a system infrastructure that provides important convenience for both process owners and internal auditors, it can also reveal biased results as explained in previous chapters. Among the prospects for the future, it is expected that artificial intelligence applications will most likely act more autonomously. Because of this, AI-based systems will make more complex decisions that require using a human judgment-based approach (Nelson and Tan 2005) beforehand.

The point to be emphasized here is to determine whether AI can be trusted before using it extensively in business processes. Recently, with the widespread use of artificial intelligence, concerns have been raised regarding data privacy, cybersecurity, and how it may affect the environment. Although traditional technologies generally work on deterministic and in other words "predictable software", artificial intelligence algorithms work differently

(Kokino and Davenport 2017). Therefore, to be able to rely on artificial intelligence, it is necessary to test, audit, and document processes by internal auditors more comprehensively.

COSO's control environment component is critical in the digital transformation process for the HE sectors. The internal control environment provides a basis for achieving high performance in a higher education institution, increasing accountability and the effectiveness of internal controls. Recently, concerns have arisen about the lack of accountability arising from the complex processes of universities. In this context, the internal audit steps in as a consultant to encourage accountability throughout the organization and to spread ethical practices.

Here, the internal audit begins to use data intelligence to provide transparency and visibility into key accountability indicators and to track them quantitatively. Thus, internal audit becomes a fundamental step toward providing real-time transparency to appropriate controls, authorization, and problem management, and even whether individuals are displaying the correct behaviors or not through CA/CM implementations (Vasarhelyi et al. 2020).

6.5 FUNCTIONAL AND ADMINISTRATIVE REPORTING LINES WITH DIGITAL AUDIT REPORTING REQUIREMENTS

Regular reporting should be one of the main responsibilities of a chief audit executive according to IIA standards and a means of conducting healthy corporate communication. The purpose, scope, and considerations of this reporting are considered as a process, and the recommendations for best practices are presented within the framework of IIA standards. Hence, compliance of the higher education sector with these reporting standards is a fundamental requirement. However, due to the complex structure of the higher education sector within the framework described with different perspectives in various chapters of this book, it is expected to establish a reporting structure beyond the IIA standards.

The main reason for this is that it is a fundamental requirement to support reporting with advanced infrastructure in terms of time and quality, as the higher education sector operates with a dynamic structure and has a mission that produces information for the whole society.

The reporting mechanism guides senior management in decision-making processes with an approach that addresses internal and external needs. In addition, the goal of creating a "role model" and building a "corporate culture" throughout the organization is important and a priority in the reporting presented to the process owners.

In this context, the internal audit department of a university should periodically report to the senior management and the board of directors about the purpose of the internal audit activity, the areas of responsibility and authority, and the performance indicators realized according to risk-based audit plans. However, it would be appropriate to submit a report to deter fraud risks (Kranacher 2019). Internal audit is required to reveal the management problems that may arise in the digital transformation process and to report important risk exposures and control problems by conducting investigations on other issues that are needed or requested by the senior management and the board of directors.

IIA Standard 1110 states that chief audit executives must report to a level that allows them to fulfill their audit and consulting responsibilities within the framework of corporate governance principles (IIA 2017a; 2017b). The main purpose here is to ensure that the top management's duties are determined according to the risk-based audit plan and that there is a level that prevents adversely affecting the results of each individual audit engagement. In the ideal framework, the internal audit should present its reports to the audit committee functionally as defined and administratively to the chief executive officer and provide information based on the 5Cs approach (Figure 6.4) in regular communication.

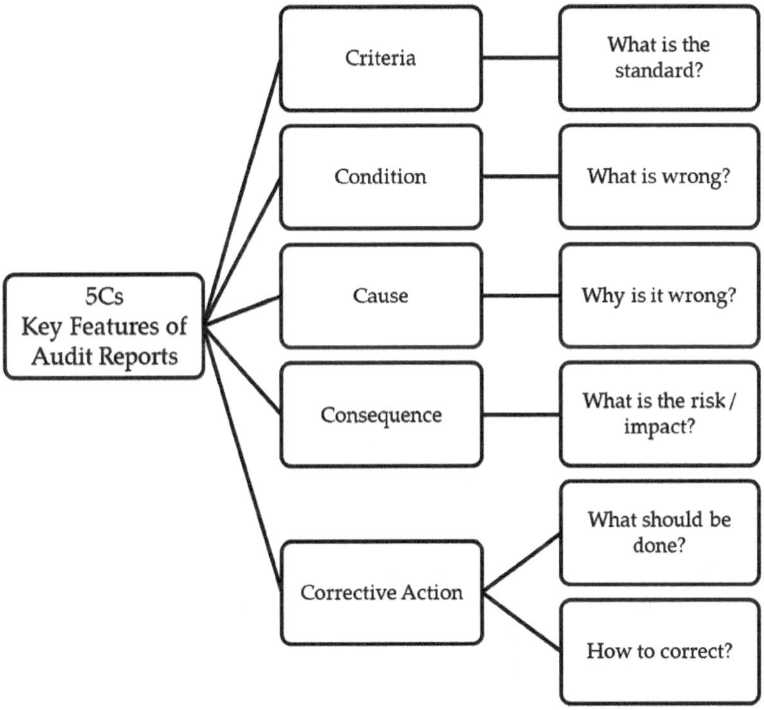

Figure 6.4 The 5Cs-key features of audit reports.

Source: WIRC of ICAI (2020).

It is a fact that the functions of the tools used in the auditing of the processes are decisive. In this respect, the relationship between the audit tools and objectives in every audit process will become the effectiveness of the functions in question. Therefore, the function of reporting in the audit process will be evaluated by how much the audit recommendations and findings show about the processes for the manager and affects the decision-making process.

This assessment will also have an impact on the effectiveness of internal audits as a risk management tool to reduce losses arising from risks. Hence, there is a need for an "integrated reporting" system (Faleato and Ki-Zerbo 2015) that takes into account all financial and non-financial business processes and possible factors within the scope of internal audit risk management for the HE institution (SDG 2017).

Public policies that are environmentally sensitive and based on a sustainable development approach force the HE institutions to be more open on this issue. In general, instead of checking whether they are complied with because of the policies put into practice by the authorities, the HE institutions audit themselves. In this way, they explain their sustainability and responsible management approach to the public openly and transparently (Bhowmik et al. 2017). As a result, the internal audit departments of the HE institutions have a new function based on integrated reporting requirements.

As a result of the determination of public policies by considering the climate and the areas that it may affect, and the best practices based on "sustainable development" understanding, which aims to leave resources, especially "green energy", to future generations, the HE institutions have to set new goals other than income or financial goals (Bhowmik et al. 2017).

This situation expands to include all other stakeholders, including the area in which the HE institutions operate. As a result, a reporting infrastructure that creates value comprising the present and the future, taking into account the long-term objectives of the HE institutions will be established (SDG 2017). In this structure, the functioning of the reporting system should be based on social accountability, especially for the prevention of shocks such as COVID-19 that can affect the environment or/and the coexistence of people.

This reporting structure will reveal an interactive and networked relationship with external stakeholders. This networked relationship makes communication a fundamental mechanism within the new audit approach (Carnevale 2019). In this context, the leading means of communication are these integrated reports. In this respect, each report will be a communication tool between sub-units and units, and between processes within the entire HE institution. This situation transforms the audit process that produces these reports into a structure that monitors the smallest units of the HE institution and transmits the information obtained from it to the top management. When evaluated from this point of view, a structure similar to the "neural network system" (Corpuz 2019) has emerged when considering the complex business processes in the new audit approach. Within this structure, as stated before, reporting can be thought of as collections of signals through which information is transmitted.

In the new reporting approach, it is considered that human and their relationship with nature constitute a "social capital" (Ciesewicz and Ciesewicz 2018). Universities are generally institutions in which social capital is included. This capital is the social structure that transfers this resource to the production processes that meet the needs of the community and society. In this respect, it explains to society how much this task is fulfilled and how functional it is in the reporting standards with the integrated reporting approach. Hence, these reports make the HE institutions accountable to society as well as overseeing their primary function for society. For this reason, this is one of the main reasons for the differentiation of university auditing today.

REFERENCES

Abraham, J.M., Walker, P.L. (2017). ERM and reputational risk: More talk than action? https://www.ue.org/uploadedFiles/ERM%20and%20Reputational%20Risk%20White%20Paper.pdf

Agasisti, T. (2016). Cost structure, productivity and efficiency of the Italian public higher education industry 2001–2011. *International Review of Applied Economics*, 30(1), 48–68. DOI: 10.1080/02692171.2015.1070130

Aksoy, T., Kahyaoglu, S. (2013). Measuring the internal audit performance: Tips for successful implementation in Turkey. *American International Journal of Contemporary Research*, 3(4), 71–82. http://www.aijcrnet.com/journals/Vol_3_No_4_April_2013/8.pdf

Allegrini, M., D'Onza, G., Paape, L., Melville, R., Sarens, G. (2006). The European literature review on internal auditing. *Managerial Auditing Journal*, 21(8), 845–853. DOI: 10.1108/02686900610703787

Allui, A. and Jolly, S.. (2016). Strategic human resource management in higher education institutions: Empirical evidence from Saudi. 12th International Strategic Management Conference, ISMC 2016, 28-30 October 2016, Antalya, Turkey. *Procedia – Social and Behavioral Sciences*, 235, 361–371.

Alshbiel, S. O. (2017). Internal auditing effectiveness success model: A study on Jordanian industrial firms. In *Proceedings of the Second American Academic research conference on global business, economics, finance and social sciences* (pp. 978–1001).

Alzeban, A. (2015). The impact of culture on the quality of internal audit. *Journal of Accounting, Auditing & Finance*, 30(1), 57–77. DOI: 10.1177/0148558X14549460

Alzeban, A., Gwilliam, D. (2014). Factors affecting the internal audit effectiveness: A survey of the Saudi public sector. *Journal of International Accounting, Auditing and Taxation*, 23(2), 74–86. DOI: 10.1016/j.intaccaudtax.2014.06.001

Arena, M., Azzone, G. (2009). Identifying organizational drivers of internal audit effectiveness. *International Journal of Auditing*, 13, 43–60. DOI: 10.1111/j.1099-1123.2008.00392.x

Badara, M. S., Saidin, S. Z. (2013). The journey so far on internal audit effectiveness: A calling for expansion. *International Journal of Academic Research in Accounting, Finance and Management Sciences*, 3(3), 340–351. DOI: 10.6007/IJARAFMS/v3-i3/225

Badara, M. S., Saidin, S. Z. (2014). Empirical evidence of antecedents of internal audit effectiveness from Nigerian perspective. *Middle-East Journal of Scientific Research*, 19(4), 460–471. DOI: 10.5829/idosi.mejsr.2014.19.4.1783

Baheri, J., Sudarmanto, Wekke, I. S. (2017). The effect of management support to effectiveness of internal audit for public universities. *Journal of Engineering and Applied Sciences*, 12(7), 1696–1700. DOI: 10.3923/jeasci.2017.1696.1700

Bailey, J. A. (2016). Looking to the future for internal audit standards. *Updates, Usage, and Conformance*. https://www.iia.nl/SiteFiles/Publicaties/Foundation%20CBOK%20Looking%20to%20the%20Future%20for%20Internal%20Audit%20Standards%20August%202016.pdf

Baker Tilly (2019). Rethinking organizational relationships to strategically align compliance, audit, risk, ethics and ERM functions in higher education. https://baker-tilly-www.cdn.prismic.io/baker-tilly-www/a3509b6f-3c46-4e90-82d0-4be8fc7ac62b_aligning-compliance-audit-risk-ethics-and-erm-in-higher-education.pdf

Baletic, B., Lisac, R., Pap, M. (2020). A transition to sustainable lifestyles the role of universities. In U. M. de Miranda Azeiteiro & J. Paulo Davim (Eds.), *Higher education and sustainability: Opportunities and challenges for achieving sustainable development goals*. Boca Raton: Taylor & Francis Publications. ISBN:9781138556539.

Bednarek, P. (2018). Factors affecting the internal audit effectiveness: A survey of the polish private and public sectors. In Tadeusz Dudycz, Grażyna Osbert-Pociecha and Bogumiła Brycz (Eds.), *Efficiency in business and economics* (pp. 1–16). Springer. DOI: 10.1007/978-3-319-68285-3_1

Beni, M. (2019). A simple revolution for digital auditing and auditing digital. https://www.icaew.com/technical/audit-and-assurance/faculty/audit-and-beyond/audit-and-beyond-2019/audit-and-beyond-december-2019/a-simple-revolution-for-digital-auditing-and-auditing-digital

Benneworth, P. and Cunha, J. (2015). Universities' contributions to social innovation: Reflections in theory & practice. *European Journal of Innovation Management* 18(4), 508–527. DOI: 10.1108/EJIM-10-2013-0099

Betti, N., Sarens, G. (2020). Understanding the internal audit function in a digitalised business environment. *Journal of Accounting & Organizational Change* 17(2), 197–216. DOI:10.1108/JAOC-11-2019-0114

Bhowmik, J., Selim, S. A., Huq, S. (2017). The role of universities in achieving the sustainable development goals. *CSD-ULAB and ICCCAD Policy Brief*. ULAB, Dhaka. http://www.icccad.net/wp-content/uploads/2015/12/Policy-Brief-on-role-of-Universities-in-achieving-SDGs.pdf

Bozkus Kahyaoglu, S. (2017). The role of internal auditors in adopting ethical culture in the public institutions in the framework of the relation between internal audit and ethics. *Technical Report*. EU Project Technical Assistance for Prevention of Corruption and Promotion of Ethics. https://isletme.bakircay.edu.tr/Yuklenenler/Resimler/iibf/resimler/Teknik_Rapor_ingilizce.pdf

Bozkus Kahyaoglu, S., Coskun, E. (2020). *Rethinking audit culture in digital transformation era: Recommendations for auditing digitally transforming universities*. In *The 11th international conference on governance fraud ethics and corporate social responsibility (CSR)* which was virtually hosted by London Metropolitan University, Guildhall School of Business and Law on 27–28 August, 2020.

Bozkus Kahyaoglu, S, Sarıkaya, R, Topal, B. (2020). Continuous auditing as a strategic tool in public sector internal audit: The turkish case. *Journal of Selçuk*

University Social Sciences Vocational School (Selçuk Üniversitesi Sosyal Bilimler Meslek Yüksekokulu Dergisi), 23(1), 208–225. DOI:10.29249/selcuksbmyd.670261

Cangemi, M. P. (2015). *Staying a step ahead internal audit's use of technology*. Florida, USA: The Institute of Internal Auditors Research Foundation (IIARF) Publications. https://www.researchgate.net/publication/281176816_Staying_a_Step_Ahead_Internal_Audit%27s_Use_of_Technology_by_Michael_P_Cangemi

Carnevale, M. (2019). Connect further: Advice for auditors new to higher education. *College and University Auditor Journal*, Summer, 1–6. https://acua.org/College-and-University-Auditor-Journal/Summer/Connect-Further-Advice-for-Auditors-New-to-Higher

Chambers, A. D., Odar, M. (2015). A new vision for internal audit. *Managerial Auditing Journal*, 30(1), 34–55 Follow journal. DOI: 10.1108/MAJ-08-2014-1073

Chandola, V. (2015). Digital Transformation and Sustainability: Study and Analysis. Harvard University, Extension Studies in candidacy for the degree of Master of Liberal Arts (ALM) in Sustainability. https://www.researchgate.net/publication/292983072_Digital_Transformation_and_Sustainability

Chartered Institute of Internal Auditors. (2020). Position paper: Internal audit's relationship with external audit. https://www.iia.org.uk/resources/delivering-internal-audit/position-paper-internal-audits-relationship-with-external-audit/

Christopher, J. (2012). The adoption of internal audit as a governance control mechanism in Australian public universities – Views from the CEOs. *Journal of Higher Education Policy and Management*, 34(5), 529–541.

Chung, J. (2020). Becoming a data-driven organization. https://d1.awsstatic.com/executive-insights/en_US/ebook-becoming-data-driven.pdf

Ciesewicz, E., Ciesewicz, J. (2018). The social capital approach. *College and University Auditor Journal*, Fall, 1–5. https://acua.org/College-and-University-Auditor-Journal/Fall-2018/The-Social-Capital-Approach

Ciolek, M. (2017). Professional skepticism in auditing and its characteristics. Global Challenges of Management Control and Reporting. ISSN 1899-3192. https://www.researchgate.net/publication/324254481_PROFESSIONAL_SKEPTICISM_IN_AUDITING_AND_ITS_CHARACTERISTICS

CohnReznick. (2020). 2020: Audit quality and transparency. https://www.cohnreznick.com/insights/audit-quality-and-transparency

Committee of Sponsoring Organizations of the Treadway Commission (COSO). (2020). Blockchain and internal control, COSO perspective. https://www.coso.org/Documents/Blockchain-and-Internal-Control-The-COSO-Perspective-Guidance.pdf

Corpuz, R. S. A. (2019). Implementation of artificial neural network using scaled conjugate gradient in ISO 9001:2015 audit findings classification. *International Journal of Recent Technology and Engineering (IJRTE)*, 8(2), 420–425. ISSN: 2277-3878. https://www.researchgate.net/publication/334762782_Implementation_of_Artificial_Neural_Network_Using_Scaled_Conjugate_Gradient_in_ISO_90012015_Audit_Findings_Classification

D'Onza, G., Selim, G. M., Melville, R., Allegrini, M. (2015). A study on internal auditor perceptions of the function ability to add value. *International Journal of Auditing*, 19(3), 182–194. DOI: 10.1111/ijau.12048

Dejnaronk, J., Little, H. T., Mujtaba, B. G., McClelland, R. (2016). Factors influencing the effectiveness of the internal audit function in Thailand. *Journal of Business and Policy Research*, 11(2), 80–93.

Dellai, H., Ali, M., Omri, B. (2016). Factors affecting the internal audit effectiveness in Tunisian organizations. *Research Journal of Finance and Accounting*, 7(16), 2222–2847.

Deloitte. (2017). Model risk management. Driving the value in modelling. https://www2.deloitte.com/content/dam/Deloitte/fr/Documents/risk/deloitte_model-risk-management_plaquette.pdf

Deloitte. (2018a). Significant risks facing higher education taking an enterprise approach to risk management. https://www2.deloitte.com/us/en/pages/public-sector/articles/higher-education-issues-and-enterprise-risk-management.html

Deloitte. (2018b). Digital maturity model achieving digital maturity to drive growth. https://www2.deloitte.com/content/dam/Deloitte/global/Documents/Technology-Media-Telecommunications/deloitte-digital-maturity-model.pdf

Deloitte. (2018c). Internal Audit 3.0 the future of internal audit is now. https://www2.deloitte.com/content/dam/Deloitte/global/Documents/Audit/gx-internal-audit-3.0-the-future-of-internal-audit-is-now.pdf

Deloitte. (2020a). Internal audit insights high-impact areas of focus – 2020. www2.deloitte.com/content/dam/Deloitte/ie/Documents/Risk/IE-risk-inernal-audit-insights-2020.pdf

Deloitte. (2020b). Modernizing the three lines of defense model an internal audit perspective. https://www2.deloitte.com/us/en/pages/advisory/articles/modernizing-the-three-lines-of-defense-model.html

DeSimone, S., Rich, K. (2019). Determinants and consequences of internal audit functions within colleges and universities. https://web.holycross.edu/RePEc/hcx/HC1904-DeSimone-Rich_IAFUniversities.pdf

Endaya, K. A., Hanefah, M. M. (2013). Internal audit effectiveness: An approach proposition to develop the theoretical framework. *Research Journal of Finance and Accounting*, 4(10), 2222–2847.

Erasmus, L. J., Coetzee, P. (2018). Drivers of stakeholders' view of internal audit effectiveness: Management versus audit committee. *Managerial Auditing Journal*, 33(1), 90–114. DOI: 10.1108/MAJ-05-2017-1558

EY. (2018). In a digital world, do you know where your risks are? Key considerations for your internal audit plan to help management navigate in the Transformative Age. https://assets.ey.com/content/dam/ey-sites/ey-com/en_gl/topics/digital/EY-In-a-digital-world-do-you-know-where-your-risks-are-sa-final.pdf

Faleato, J., Ki-Zerbo, B. (2015). *Enhancing integrated reporting internal audit value proposition*. European Institutes of Internal Auditors (IIA France, IIA Netherlands, IIA Norway, IIA Spain, IIA UK and Ireland) with the contribution of IIA Global and the IIRC. ISBN:978-84-943299-1-3. https://integratedreporting.org/resource/enhancing-integrated-reporting-internal-audit-value-proposition/

FINEEC – Finnish Education Evaluation Centre. (2017). Audit Manual for Higher Education Institutions 2018–2024. Publications 21:2017. ISBN:978-952-206-412-7 (pdf). https://karvi.fi/app/uploads/2017/02/FINEEC_Audit_manual_for_higher_education_institutions_2018-2024_FINAL.pdf

Fonseca, A. R., Jorge, S., Nascimento, C. (2020). The role of internal auditing in promoting accountability in higher education institutions. *Journal of Public Administration*, 54(2), 243–265. DOI: 10.1590/0034-761220190267x

Grant Thornton. (2017). Corporate risk frameworks. Driving business strategy with effective risk frameworks. https://www.grantthornton.co.uk/globalassets/1.-member-firms/united-kingdom/pdf/documents/corporate-risk-frameworks.pdf

Green Book. (2017). Enhancing university board governance and effectiveness university. *Transformation Programme, Green Book*. Malaysia: Ministry of Higher Education.

Halimah, N. A., Othman, R., Othman, R., Jusoff, K. (2009). The effectiveness of internal audit in Malaysian public sector. *Journal of Modern Accounting and Auditing*, 5(9). ISSN 1548-6583.

IIA. (2010). Measuring Internal Audit Effectiveness and Efficiency. IPPF Practical Guide. Institute of Internal Auditors (IIA) Report.

IIA. (2016). Measuring the Effectiveness of the Internal Audit Function. Practical Tools for Internal Auditors. Institute of Internal Auditors (IIA) Report.

IIA. (2017a). International standards for the professional practice of internal auditing. Institute of Internal Auditors (IIA) Report.

IIA. (2017b). Internal audit and external audit distinctive roles in organizational governance. *Global Perspectives and Insights* 8, 1–7. https://www.iia.org.uk/media/1691032/distinctive-roles-in-organisational-governance.pdf

IIA. (2020). International Professional Practices Framework (IPPF). The Institute of Internal Auditors, Altamonte Springs. https://na.theiia.org/standards-guidance/Pages/Standards-and-Guidance-IPPF.aspx

Jiang, L., Messier, Jr. W.F., Wood, D.A. (2016) The association between internal audit operations-related services and firm operating performance. *Auditing: A Journal of Practice & Theory* 39(1), 101–124. DOI: 10.2308/ajpt-52565

Jingura, R. M., & Kamusoko, R. (2019). A competency framework for internal quality assurance in higher education. *International Journal of Management in Education* 13(2), 119–132.

Kane, G.C., Palmer, D., Nguyen Phillips, A., Kiron, D., Buckley, N. (2016). Aligning the organization for its digital future. *MIT Sloan Management Review*, 58(1), 1–27.

Kokino, J., Davenport, T. H. (2017). The emergence of artificial intelligence: how automation is changing auditing. *Journal of Emerging Technologies in Accounting*, 14(1), 115–122. DOI: 10.2308/jeta-51730

KPMG. (2020a). 20 key risks to consider by Internal Audit before 2020. Are you aware of the risks concerning Internal Audit today and in the near future? https://assets.kpmg/content/dam/kpmg/ch/pdf/key-risks-internal-audit-2018.pdf

KPMG. (2020b). On the 2020 Higher Education Audit Committee Agenda. https://institutes.kpmg.us/content/dam/institutes/en/government/pdfs/2020/2020-he-audit-cmte.pdf

KPMG Audit Committee Institute. (2016). Effectiveness of the internal audit function. Audit Committee Questions. https://assets.kpmg/content/dam/kpmg/pdf/2016/07/3-aci-assessing-internal-effectiveness-fs-uk-v5-lr-(002).pdf

Kranacher, M.-J. (2019). Fighting fraud and corruption in higher education. https://assets.corporatecompliance.org/Portals/1/PDF/Resources/past_handouts/Higher_Ed/2019/602_Fighting%20Fraud%20and%20Corruption.pdf

Lenz, R., Hahn, U. (2015). A synthesis of empirical internal audit effectiveness literature pointing to new research opportunities. *Managerial Auditing Journal*, 30(1), 5–33. DOI: 10.1108/MAJ-08-2014-1072

Lenz, R., Sarens, G., Hoos, F. (2017). Internal audit effectiveness: multiple case study research involving chief audit executives and senior management. *EDPACS, The EDP Audit, Control, and Security Newsletter*, 55(1), 1–17. DOI: 10.1080/07366981.2017.1278980

Lenz, R., Sarens, G., Jeppesen, K. K. (2018). In search of a measure of effectiveness for internal audit functions: an institutional perspective. *The EDP Audit, Control, and Security Newsletter (EDPACS)*, 58(2), 1–36. DOI: 10.1080/07366981.2018.1511324

Li, F., Chen, X. (2012). Economies of scope in distance education: The case of Chinese research universities. *The International Review of Research in Open and Distributed Learning*, 13(3), 117–131.

Liang-Cheng, Z. (2016). Economies of Scale and Scope in Higher Education. PhD Dissertation. Griffith Business School. Department of Accounting, Finance and Economics. https://researchrepository.griffith.edu.au/bitstream/handle/10072/366522/Zhang,%20Liang-Cheng_Final%20Thesis_Redacted.pdf?isAllowed=y&sequence=1

Lindberg, S. (2013). Mapping accountability: core concept and subtypes. *International Review of Administrative Sciences*, 79(2), 202–226.

Marques, M. C. C. (2017). Corporate governance and international audit in public higher education institutions in Portugal. *Revista Científica Guillermo de Ockham*, 15(1). https://www.redalyc.org/journal/1053/105352363011/html/

Martel-Escobar, M., Francisco-José, V.-P., Agustín, H.-B. (2018). Bayesian inference in auditing with partial prior information using maximum entropy priors. *Entropy*, 20, 919. DOI: 10.3390/e20120919. https://www.mdpi.com/1099-4300/20/12/919/htm

Master, K., Bansal, S. (2019). Impact of digital on the future of internal audit. EXL White Paper. https://www.exlservice.com/resources/assets/library/documents/EXL-WP-Impact-of-Digital-on-the-Future-of-Internal-Audit.pdf

McKinsey. (2017a). Digital risk: Transforming risk management for the 2020s. https://www.mckinsey.com/business-functions/risk/our-insights/digital-risk-transforming-risk-management-for-the-2020s

McKinsey. (2017b). Culture for a digital age. https://www.mckinsey.com/business-functions/mckinsey-digital/our-insights/culture-for-a-digital-age

Mihret, D. G., Yismaw, A. W. (2007). Internal audit effectiveness: An Ethiopian public sector case study. *Managerial Auditing Journal*, 22(5), 470–484. DOI: 10.1108/02686900710750757

Nelson, M., Tan, H. T. (2005). Judgment and decision making research in auditing: A task, person, and interpersonal interaction perspective. *Auditing A Journal of Practice & Theory*, 24(s-1). DOI: 10.2139/ssrn.761706. https://www.researchgate.net/publication/228182837_Judgment_and_Decision_Making_Research_in_Auditing_A_Task_Person_and_Interpersonal_Interaction_Perspective

Newman, C. (2017). Data Analytics is Driving Efficiency in Higher Ed. *The College and University Auditor Journal* Fall, 1–6. https://acua.org/College-and-University-Auditor-Journal/Fall-2017/Data-Analytics-is-Driving-Efficiency-in-Higher-Ed

Ntim, C. G., Soobaroyen, T., Broad, M. J. (2017). Governance structures, voluntary disclosures and public accountability: The case of UK higher education institutions. *Accounting, Auditing and Accountability Journal*, 30(1), 65–118.

Nurdiono, Gamayuni, R. R. (2018). The effect of internal auditor competency on internal audit quality and its implication on the accountability of local government. *European Research Studies Journal*, 21(4), 426–434.

Oliver Wyman. (2020). Right sizing of the three lines of defense – How to make risk management work for you. https://www.oliverwyman.com/content/dam/oliver-wyman/v2/publications/2020/January/Oliver_Wyman_Right_Sizing_The_Three_Lines_Of_Defense_paper.pdf

Origami Risk. (2020). Facing the challenge of reputation management in higher education. https://www.origamirisk.com/sites/default/files/file/2020-09/Challenge-of-Reputation-Management-Higher-Ed-EB.pdf

Oussii, A. A., Taktak, N. B. (2018). The impact of internal audit function characteristics on internal control quality. *Managerial Auditing Journal*, 33(5), 450–469.

Peterson, B. G., Carl, P. (2014). Performance Analytics: Econometric tools for performance and risk analysis. Retrieved from http://CRAN.R-project.org/package=PerformanceAnalytics

Pitt, S. A. (2014). *Internal audit quality developing a quality assurance and improvement program*. Wiley Publications. ISBN:978-1-118-71550-5 (ePDF).

PwC. (2006). *Achieving goals and protecting reputation*. ERM for Educational Institutions. https://regents.universityofcalifornia.edu/regmeet/july08/a7a.pdf

PwC. (2014). Perspectives in Higher Education. https://www.pwc.co.za/en/assets/pdf/perspectives-in-higher-education-2014.pdf

PwC (2015). State of the internal audit profession study: Finding true North in a period of rapid transformation. Available at: https://www.pwc.com/ve/es/auditoria-interna/assets/pwcs-2015- state-of-the-internal-audit-profession-study.pdf

PwC. (2018). State of the internal audit profession study. Moving at the speed of innovation: The foundational tools and talents of technology-enabled internal audit. Available at: https://www.pwc.com/sg/en/publications/assets/state-of-the-internal-audit-2018.pdf

PwC, ACCA, INSEAD. (2019). Re-inventing internal controls in the digital age. https://www.pwc.com/sg/en/publications/assets/reinventing-internal-controls-in-the-digital-age-201904.pdf

Ross, S. (2019). Internal vs. external economies of scale: What's the difference? https://www.investopedia.com/ask/answers/013015/what-are-differences-between-internal-and-external-economies-scale.asp

RSM. (2020). The IIA's three lines model (an update of the three lines of defense). https://www.rsm.global/australia/insights/technology-and-risks/iias-three-lines-model-update-three-lines-defense

Sarens, G., Mohammad, J. A., Rainer, L. (2012). Factors associated with the internal audit function's role in corporate governance. *Journal of Applied Accounting Research*, 13(2), 191–204. Emerald Group Publishing Limited 0967-5426. DOI: 10.1108/09675421211254876

SDG Knowledge Hub Reports. (2017): Focus on role of universities in achieving SDG. http://sdg.iisd.org/news/reports-focus-on-role-of-universities-inachieving-sdgs/

Struthers-Kennedy, A., Fitzgerald C. (2019). *Bridging the divide: Auditing digital risk*. Protiviti. https://blog.protiviti.com/2019/05/02/bridging-the-divide-auditing-digital-risk/

Türetken, S., Jethefer S., Ozkan, B. (2020). Internal audit effectiveness: Operationalization and influencing factors. *Managerial Auditing Journal*, 35(2), 238–271. DOI: 10.1108/MAJ-08-2018-1980

Turner, D.A. (2011). *Quality in higher education*. Sense Publishers. ISBN: 978-94-6091-684-7 (e-book).

Vasarhelyi, M. A., Cho, S., Cheong, A., Chanyuan (Abigail), Z. (2020). Smart audit: The digital transformation of audit. *Journal of EU Court of Auditors*, 1, 27–33. Big Data & Digital Audit. No. 1. 2020.

Western India Regional Council of Institute of Chartered Accountants of India (WIRC of ICAI). (2020). Presentation On Internal Audit Reporting. https://www.wirc-icai.org/images/material/Internal-Audit-Reporting-SP.pdf

Wise. (2018). Assessing the use of geospatial technologies in higher education teaching. *European Journal of Geography*, 9(3), 1154–1164. https://www.researchgate.net/publication/330689449_Assessing_the_use_of_geospatial_technologies_in_higher_education_teaching

Yee, C. S. L., Sujan, A., James, K., Leung, J. K. S. (2008). Perceptions of Singaporean internal audit customers regarding the role and effectiveness of internal audit. *Asian Journal of Business and Accounting*, 1(2), 147–174.

Zipper, J. (2016). Maximize value, adopt a flexible approach to auditing major projects. https://www.isaca.org/resources/isaca-journal/issues/2016/volume-3/maximize-value-adopt-a-flexible-approach-to-auditing-major-projects

Chapter 7

The impact of the pandemic crisis on higher education institutions with post-COVID auditing perspective

7.1 THE IMPACT OF DIGITAL TRANSFORMATION ON THE HIGHER EDUCATION SYSTEM

Digital transformation essentially refers to the application of technology to create new business models, processes, software, and systems for organizations to gain more returns, provide better competitive advantage, and more efficiency (CISCO 2016). In general, organizations can achieve this by mainly transforming business models, generating efficiency and innovation, and customizing stakeholder experiences (Schallmo et al. 2017; Schwertner 2017).

According to Otola and Grabowska (2020), a digital institution refers to an organization that benefits from data and data analytics. In fact, data are available in analog work as well as in digital. However, the point to be considered here is that it becomes riskier and more valuable in the digital environment. This approach lays the foundation for analytics, new, smart, and advanced work style in the HE sectors. Even if an HE institution has not yet decided to fully adopt a digital business platform, big data and analytics need to be explored to improve core business outcomes and position big data and analytics at the center of their digitalization strategy (Deloitte 2020).

It is a fact that digital transformation in the world is accepted as a subject that is already widely adopted and included in the agenda of various institutions for the implementation process. However, with the COVID-19 pandemic that has emerged recently, it appears as the only and, perhaps the top priority issue for the institutions (McKinsey 2020).

With COVID-19, we face a public health crisis caused by such a far-reaching mass deterioration (Di Pietro et al. 2020). This situation brings about a global economic recession. In this context, students' university preferences and enrollment rates remain uncertain. The concerns of both students and parents are constantly increasing. Despite this, managements of higher education institutions must act as leaders who show their broad perspectives to respond accordingly. That is needed to direct society and take the necessary strategic decisions without delay (Otola and Grabowska 2020).

DOI: 10.1201/9781003093008-7

With the recent pandemic, higher education has no choice but to invest in technology in the face of limited public funding and high social and economic expectations for what can be achieved. One way to maximize the use of more technology to meet the challenges here is to foster collaboration on a large scale, taking into account "economies of scale" and shared technological risks. This approach can reduce repetition and wasted competition between universities (Koshal and Koshal 1995; Johnes and Johnes 2016; Toutkoushian 2016; Rogoff 2020). The relevant point here is that the cooperation in question can be between public institutions and may include private institutions. Collaboration and risk sharing can also be applied in institutions that are large practice and research centers, for example, between different university departments (Di Pietro et al. 2020).

Above all, higher education leaders must make decisions to determine how to balance the financial consequences of this profound crisis, student experience and safety, and equal access to high-quality education. It should be noted that the McKinsey Research Report (2020) provides valuable information about the dramatic change in the HE sectors based on survey findings. Namely, there is a change in students' university preferences due to COVID-19 and this is given in Figure 7.1.

With the distance education of the Higher Education Institutions during the pandemic period, doubts began to arise about whether the education in social sciences and non-laboratory-based areas will lead to a change in future career and living standards (Aristovnik et al. 2020).

Developments toward the disappearance of traditional professions push young people to new preferences. As a result, the demand for traditional universities has started to decline (Jin et al. 2015). The expectation of young people from the universities is primarily associated with the job opportunity they will provide upon graduation. However, when the income level of individuals who do not have a very high level of education and their prominence in society is associated with the above result, there is a negative effect on the demand for the universities (Li 2013).

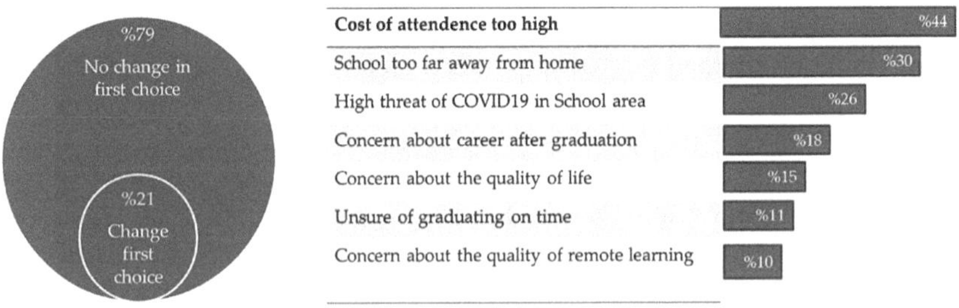

Figure 7.1 COVID-19 higher education enrollment survey – key findings.

Source: McKinsey (2020).

The fact that the necessary information in life can be obtained from different online sources leads to an effect that will reduce the demand for universities outside of basic sciences and laboratory-based sciences. In this respect, an imperative search will emerge for what needs to be accomplished for sure. First of all, a decrease in demand will cause universities to face a financial problem. This reduction may lead to new resource problems regarding investments in technological infrastructures. As a result, with the decrease in the welfare level of the experienced and talented faculty members, new quests will arise (Cardoso et al. 2007; Hamzaee 2015; UNESCO International Institute for Higher Education (IESALC) 2020). However, with the increase of interdisciplinary interaction and transmittance, the necessity for individuals who are equipped for new professions should increase and universities should undergo a rapid structuring in these areas. In this sense, the dissemination of education will be put into effect as a best practice. That will provide the HE institutions to establish the fields with an interdisciplinary perspective and direct information to innovations, especially in new education systems (Davies and Devlin 2010; Pramanik 2014; Ribeiro and Relvas 2018).

Technological developments and the widespread use of online education are some of the driving forces of change in higher education. However, this is one of the responses to other changes. In this context, technological advances offer very important and innovative opportunities to adapt learning and teaching to the mass higher education environment. It can be considered as a basic tool to improve the "quality of the learner's experience", which is among the strategic goals of higher education institutions (Tan et al. 2016).

The most important effect of the pandemic in the HE institutions is the emergence of health, especially public health, as a risk factor within communities (UNESCO IESALC 2020). The evaluations of universities, which have student resources especially at the international level and from a wide variety of countries, show that they rank second after cybersecurity (Kim 2020). Beyond that the pandemic affects only one institution, the results are the same in each HE institution; it is necessary to establish a new approach toward establishing cooperation and common standards in this regard. The information produced by the internal audit department of each HE institution on this subject will constitute a source for the formation of these standards. In today's conditions, there will be a tendency that it will rank first as an important variable of health continuity in the universities. This situation may also reveal a restructure (Figure 7.2) where all members of the HE institution will follow preventive health services, especially vaccination, to reopen the door either physically, virtually, or hybrid.

Recently, it is seen that there is an increasing demand for "lifelong learning" as a new important field and target group of the services provided by the HE institutions. In a globally competitive environment, vocational training needs for a business model based on automation systems and advanced

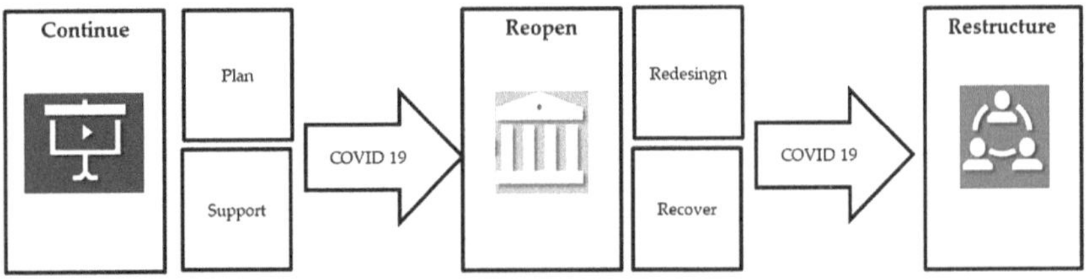

Figure 7.2 Exit framework from COVID-19 in the higher education.

Source: UNESCO IESALC (2020).

technologies are increasing (UNESCO IESALC 2019). In this context, it has been determined that new learning needs arise in areas such as the creation of general skills for employability, vocational training, and personal development.

It should be taken into account that in the digital business environment we are in, the exponential increase in knowledge, the trend toward continuous innovation in employment, and online learning opportunities will continue in the future. In this process, traditional universities and open universities in the sector play different roles (Jackson and Cooper 2012). The main reason for this is the use of targets and different approaches for different market segments.

It can be stated that the existing universities will continue to dominate undergraduate education for adults and their role in continuing education will continue to increase. From this point of view, reviewing the skill sets of higher education institutions depending on and adapting to the changing needs locally and nationally will determine their future applicability and competitiveness (Ehlers 2020).

It should be noted that it is decisive to take into account the current university education's response to employment and people's future concerns. One of the important points here is the situation of those who receive immigration from different countries. As a result of these migrations, these immigrants must participate in education, so this situation should be evaluated in another dimension (Haapanen and Böckerman 2013; Bunar 2017).

Individuals choose their professions that will determine their future level of well-being through the higher education system. Therefore, education and preference taken in the higher education system are to choose the life level in the future from today. However, such a decision made based on today's knowledge may not meet the expectations of the future. Therefore, developments that will affect occupational diversity are the main determinants of this (OECD 2008). Because technological advancements are developments leading to a change in the existing professions and the disappearance of these professions, the obligation of universities becomes responding to this situation. As a result, universities need to identify possible professions and formulate forward-looking strategies by taking into account the factors

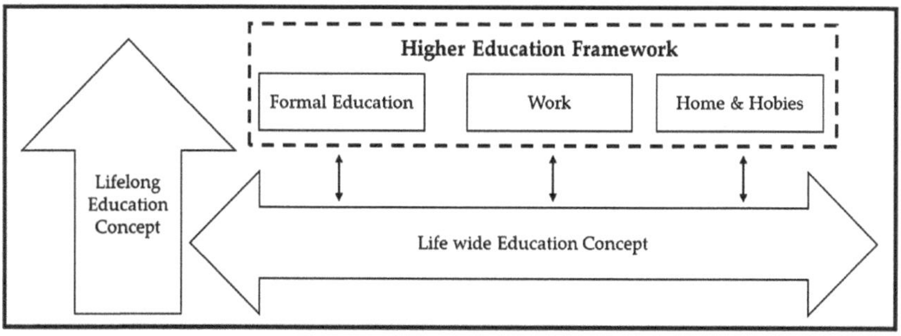

Figure 7.3 Lifelong and life wide education.

Source: Adapted from Jackson and Cooper (2012).

that lead to the development and diversification of new business fields and science branches (Ehlers 2020).

While the differentiation here reveals an intertemporal decision-making process for the future, it should encompass the present, including capacity, human resources, education, work relations, and family relations, and the talents and hobbies of the individuals. Jackson and Cooper (2012) define this approach as "life wide education" and "lifelong-education", which is given in Figure 7.3.

The understanding covers the present and the future from now; at the higher education level, there will be an understanding that includes not only the present societies but also the future generations and aims at their welfare. This understanding, within the framework of the concept of sustainable development, is a necessity to make the main purpose of higher education a process that will ensure sustainable learning and talent acquisition for the future (Kane et al. 2016; Caird and Roy 2019).

7.2 LEARNING ENVIRONMENT IN THE DIGITAL ERA

The importance of learning environments is to determine the context in which learning takes place (OECD 2020). It is a holistic concept that requires considering the elements of the environment as well as the learning that takes place within the framework of the opportunities offered by a learning environment. Therefore, it is considered as an eco-system that includes the activities and results of learning. These elements form the basis and are at the center of the learning environment.

- Learners (who?)
- Educators (with whom?)
- Content (what?)
- Learning resources (with what?), including the field

The dynamics in which these elements as a whole are critical, namely, how learners and teachers are grouped, how learning is planned and timed, and what approaches are used to learning and evaluation are also important. However, factors such as appropriate platforms, student–teacher ratios, curriculum content, pedagogical methods, and parental involvement contribute to shaping the outcomes of education (Nusche 2008; Alexander 2018).

Advances in digitalization and communication technology reveal new business processes defined as "online". However, considering the pre-COVID-19 period, the slow transformation in human resources and infrastructure, and the costly investments required for this were the most important reasons for the slowness of the digital transformation process (Filho et al. 2018).

In general, a virtual environment leads to a new type of relations and networks are emerging. In this context, especially online education sites for teaching new technologies are gaining importance apart from the courses given by universities. The increased use of information technology in higher education will help speed up and further facilitate interaction. In this context, academics can communicate extensively on innovative teaching and research projects and become more productive by working together. Beyond physically meeting on campuses, virtual communication via e-mails and online discussion platforms facilitates this. Therefore, the need for continental travel of academicians is decreasing. In particular, intercontinental cooperation will increase in the coming years as e-mails, video conferencing, online discussion platforms and other advanced technologies make this easier (Kiselova et al. 2020).

Especially web-based teaching platforms such as Coursera and Udemy transform into structures that support universities' interdisciplinary service provision. In this context, digitalization has made them influential tools for the people who want to renew their skill set and themselves. At the same time, it has created the opportunity for university students who study in a discipline to benefit from different disciplines (Bentaa et al. 2014; Vasarhelyi et al. 2020).

This situation brought along a significant acceleration in research and data analytics. However, along with the advantages of online training, there are also disadvantages. The most important of these disadvantages is the disappearance of social capital (Teijeiro-Álvarez et al. 2016). This situation connects the trust factor in business processes to the control over the online system. For this reason, it is seen that the campus and the social opportunities that universities have provided gain importance with online education in the process of COVID-19, far beyond education. However, we are in a process that occurs because the distance education tools provided by online education do not provide virtual classroom environments. In this sense, almost none of the infrastructures required for remote education are ready (Ruso et al. 2015). In this regard, there is no significant investment so far. The reason for this is that the liquidity flows caused by the economic contraction in the current COVID-19 process have created an environment that cannot meet the existing costs.

From this point of view, it is expected that the audit fieldworks of universities in the next period will reveal knowledge that can analyze the situation in this process. Therefore, the advisory service of internal audit departments will be more prominent in the next period (Reis Fonseca et al. 2020). In addition, they will contribute to the foundation of standards for the establishment of an effective online system as a result of the evaluation and analysis of the information coming from the audit fieldworks.

After the COVID-19 pandemic period, it is a fact that Higher education is no longer an opportunity only for the cream of society or for those that live in advanced countries. Most of the colleges and universities around the world are opening their doors via the Internet, and similarly, countries that have previously resisted the idea of HE and e-learning are giving in to global economic pressure (Rapaccioli 2019). In other words, they have started to accept the importance of life-long education and open particularly business schools.[1] HE sectors around the world are aiming to mass-produce internationally trained professionals to meet the demands of an increasingly global business environment based on network organizations and platforms. In this respect, market-driven, scaled-down universities that focus on practical digital business skills are gaining importance and enrollment (Tomlinson 2015, 2016).

Considering the major market players in the HE sectors, nonprofit public and private universities in the United States dominate (National Association of Independent Colleges and Universities-NAICU 2020). On the other hand, some recent players entered the fray in the HE sectors. As more and more countries proceed toward a market economy structure, the expansion of the HE sector is usually the major tenet of their reforming attempts. As a result, more universities are opening in various regions such as Asia and Eastern and Western Europe. For instance, in Asia, universities are struggling to rebuild their financial foundation after the economic reforms in the area. In addition, there are some new entrances of both for-profit entities and distance learning initiatives within the HE sectors. These HE institutions are mostly based in the United States, with some major players emerging from Western Europe.

The Bologna agreement implemented between the European Union (EU) countries and the international cooperation between higher education institutions will increase rapidly in the coming period and lead to significant changes in the sector (EU 2020). The process that started with this agreement signed by the European Ministers of Education in 1990 is developing rapidly. The point is that it pledges a comparable undergraduate and postgraduate degree framework with increasing opportunities for courses to be shared between institutions in member countries and therefore across Europe. As a critical issue in terms of standardization in quality of education, this infrastructure has been provided through the European Credit Transfer System (ECTS).

As businesses around the world request high-tech workers, community and technical colleges are no longer the only market players in the HE sector. Private firms are more than happy to step in and educate workers – for a profit. For-profit education organizations and distance learning are the new competitive favorites for students. For-profit education organizations have huge resources and hence, they can conduct their businesses internationally thanks to the Internet (Pianko and Jarrett 2012). It is a fact that off-campus learning is among the waves shaping the global higher education landscape. With technical advances and fast digitalization, HE institutions can offer education to executives, working adults, and others who do not have the time, money, or inclination to commit to the four-year, on-campus degree track (Gallagher and Palmer 2020).

The relationship between government and higher education is changing in various regions of the world, leading to the strengthening of the autonomy of HE institutions (European University Association, EUA 2020). Based on the advantages arising from autonomy, HE institutions are adopting more business-oriented personnel policies and greater market orientation. The key issue here is that "how does the higher education sector pay for new distance learning initiatives, much less the standard brick-and-mortar operations?" Due to the fundraising difficulties, consolidation is expected to accelerate in both the traditional brick and mortar universities and the for-profit sector, particularly in the distance learning sector. In this context, most of the colleges choose to cut a variety of administrative, maintenance, and recruiting costs by merging (European Foundation for Management Development, EFMD 2020). As a result, most mergers, especially in the non-profit education sector, where they prefer the term "joint affiliation", are hardly competitors.

In an economy, individuals who lose their jobs at the bottom of the business cycle start a new search. This search is a serious factor that will determine the speed of turning back from the bottom of the business cycle. Innovations and the new environment lead to the growth in the business cycle through the adaptation of individuals to these new conditions in the economy. In this respect, each economic crisis causes well-educated individuals to demand learning new things as well. Therefore, universities are the only institutions that determine the pace of exit from economic cycles (Carpentier 2020). Especially, this situation will be more determinant in today's world where the digital and digitalization-based industrial production structures develop.

What is important here and should be considered as an area of added value is that the quality of universities is in an indispensable position to provide social and economic benefits. Recently, the environmental factor has also been included in economic and social benefits within the framework of sustainability approaches. In the face of this expanding scope of quality, expectations from universities are increasing even more (Fadeeva et al. 2014). Therefore, high performance and quality are not considered a luxury, but a primary requirement. The important thing is to ensure that the standards are up-to-date and compatible with achieving strategic goals. At this point, the internal audit can develop recommendations on many critical issues and processes that can contribute to improving the quality of universities.

Today, the interconnection between higher education institutions and society is getting deeper and deeper. Besides technological developments, the "lifelong learning" philosophy can be stated as an important reason for this (Jackson and Cooper 2012). In this context, spending on the higher education sector is closely monitored by the public. At the same time, demands and expectations from higher education institutions are increasing. As a result, it requires transparent measurement of the value, impact, and benefit of higher education institutions to ensure both social and public accountability (Hevia and Vergara-Lope 2019).

Despite commitments to have more autonomy, there is growing evidence of government steering and regulation of the higher education and research system. In addition, the increasing pressure of the global rankings leads to the higher education sectors being more competitive and becoming "multipolar" while drawing the world's attention to inequalities (Hazelkorn 2013).

Today, the benefits expectations of society from higher education institutions are generally based on real problems encountered in real life. It includes a real-time demonstration of real solutions such as climate change, human health, food, water security, energy economy, or sustainable cities. This situation changes the working systematics of universities, and due to the scales and complexities of these real problems, basic social and economic problems go beyond the disciplines and require multidisciplinary approaches based on new methods and organizational structures (Sivagurunathan 2012).

Recent technological developments and transformations have demonstrated that online learning is not just about distance education. In other words, the domain of online learning is much overarching. In this context, academic practices are increasingly moving toward blended learning (also known as hybrid learning or enhanced learning) and focus on this issue in their strategic priorities (Alammary et al. 2014). Thus, it combines online learning with supportive activities on a face-to-face basis in the classroom. One of the most basic features of university education is that it requires self-learning. Viewed from this perspective, online resources and processes can help make further progress for learners in this challenging process. This approach is an opportunity owned by most of the population in many developed countries, especially within the framework of lifelong learning practices.

However, it is seen that mass higher education understanding does not reduce the competition chance of new entrants to the higher education sector. In this context, it is one of the paradoxes of the Internet and online communication that the local and the global approaches are still close to each other (Rumbley and Altbach 2016). For example, when the behaviors of individuals are examined; They tend to share with someone in the same office or community via e-mails or other virtual communication tools as well as sending e-mails to faraway points of the world. From this point of view, developing technology is not only about providing distance learning opportunities but also offering opportunities for learning in a new way that previous generations did not have.

Marton and Saljo (1997) developed the concept of "deep and surface-level processing". In this respect, "surface-level processing" is defined as the learner undertaking superficial activities such as memorizing facts. On the other hand, "deep-level processing" is defined as a concern to understand the material being studied (Biggs 1999). This means that deep learning is achieved by constructing and internalizing a deeper personal meaning. Hence, deep learning is highly appropriate when it increases the learner's ability to transfer learning for application in different cases. It is a fact that learners need to memorize some information. They use "surface-level strategies" to manage such situations. For this reason, there are some concerns about using the Internet for learning and teaching. Although the Internet has provided an enormous amount of instantly available information, it does not oversee deep understanding and the deeper levels of learning.

Today, to ensure the sustainability of the information society, it is aimed to use the resources effectively within the framework of knowledge, skills, and trends toward the learning environment. In this context, essential resources are expressed as "financial resources", "human resources", "material resources", and "time resources". With digitalization, the need for higher-level thinking skills increases in the higher education system. In particular, it is increasingly integrated with the expectations of today's and tomorrow's universities (University-Industry Innovation Network, UIIN 2020).

With digitalization, education reveals a new understanding beyond a period in which only educators transfer information, whereas students receive information. In the emergence of this understanding, it is possible to obtain information over the Internet with a visual and experience-based communication process, beyond obtaining information from a single place and a single institution (Alves and Harsh 2018). In this respect, while interdisciplinary interaction increases, the necessity and difficulty of obtaining and using the knowledge of different disciplines arise. As a result, teaching and learning will be the result of the new era as a fundamental philosophy. In this respect, the process of teaching students how to use information, how to think, and most importantly, how to do this as an expert will be fundamental.

From this point of view, it is understood that a skill-based teaching system comes to the fore with interactive learning and critical thinking. In this sense, faculty members in higher education will have the role of guiding and generally advisors and mentors. The most important problem here is how learning outcomes will be measured. This will lead to changes in the criteria for evaluation (OECD 2014).

Among these criteria, scoring-based evaluations will likely be replaced by systems such as "successful" or "unsuitable", or not. When evaluated from this point of view, it is necessary to check and measure whether the new criteria and results to ensure whether this will be an indicator as an output (Goos and Salomons 2017). It should be noted that this assessment is a process analysis. As a result, with the new understanding, internal audit emerges as a unit that can make these evaluations.

According to the uncertain and indiscrete analysis results in the pandemic, it is stated that the success of distance education is low (Leontyeva 2018). These results show that although distance education is a tool that enables interdisciplinary learning and the use of different external resources, it does not provide the effectiveness of the classroom-based education system.

These results provide information that digitalization will affect the use of course materials in classroom activities. In this respect, it is a prerequisite to redefine the classes such that digital and online education as established considering processes and tools. The most important element here will be the sharing process of course materials. The availability of information and resources or other materials related to the course in a digital environment and sharing them before the lesson can provide flexibility to students in terms of learning, both in terms of time and space. Digitalization is in front of us as a factor that can ensure the globalization of classrooms regardless of time and space (Nortvig 2014).

The emerging new process necessitates a research-based teaching system (Toom et al. 2010; Ion et al. 2014). This approach will cause individuals to tend toward learning throughout their lives. While the main purpose of this structure is to reveal the structure of a learner who can set up a problem and solve the problem, it will reveal the individual who produces solutions to the problems that are tried to be answered or solved. Within the framework of the interaction of many factors discussed, it will reveal that the study areas of the students are interdisciplinary places that allow social communication.

Education through e-learning forces the traditional structure of educational institutions and their presentation methodologies to change by paving the way for a new learning model that brings individuals to their place and home. Virtual campus applications have become widespread, and beyond distance education institutions, organizations that have regular personal communication with students have started to develop in the HE institutions where professional cultures and classroom-based curricula are the basis (DeVaney et al. 2020).

In this respect, new learning paradigms are developing as a result of online education programs. Thus, new teacher roles and functions and new didactic methodologies are developing rapidly with them (Watson and Watson 2014). We are entering a period where higher education institutions and learners need to adapt their course management and learning methodologies to the needs of these emerging innovative and technology-based forms.

It is seen that two main trends that affect the application of online learning in general in the higher education sector emerge. The first of these is campus-based learning. The second is to combine online learning into a framework of market competition or cooperation between universities and between universities and private companies. It is seen that well-prepared educational material developed for VLEs is becoming widespread in higher education institutions. It is also possible to use for campus-based students after regular face-to-face lectures and seminars (Institute for Teaching and Learning Innovation-ITaLI 2015).

In the coming years, a distance education model based on different virtual learning arrangements combined with online education databases prepared using high-quality interactive multimedia materials can be developed. In addition to these materials, universities will become a communication platform where various support functions such as face-to-face lectures and seminars, video conferences, online virtual seminars, and computer conferencing are offered for different learner groups.

The new education system is heading toward a new structuring as a process of transferring the "self-regulating learning" method, approaches, and techniques to its participants (Cassidy 2011). As a result, it will become a necessity to adopt these approaches that will ensure the planning, monitoring, and control of individuals in education systems. Thus, learning and comprehension become a competence and skill area and the relevant field of education. The reason why learning comes to the fore here is that the flexibility of the individual in employment in his/her life will be provided both for himself and the field he will work. Flexibility will increase individuals' ability to adapt to the labor market and to new areas (UNESCO International Institute for Educational Planning-IIEP 2020). Therefore, social and other economic conditions, including flexible working conditions that arise, require flexibility and, consequently, increased learning ability. Teaching how to learn is a result of a life-long learning approach.

In the digital university environment, internal auditors are expected to process complex information and produce meaningful knowledge via audit field work. At the same time, they need to have the following qualifications: systematic and critical thinking; solving problems; making decisions and judgment when considering various types of evidence; asking meaningful questions to auditees; increasing digital literacy competence; being creative, adaptable, and flexible; and they need to form a "role model" in matters such as taking responsibility for their lifelong learning (OECD Institutional Management in Higher Education, IMHE 2012).

It has been recently discussed that traditional education models are insufficient to develop the digital transformation competencies listed here. In this context, it is critical to creating innovative learning environments necessary for the competence of internal auditors.

7.3 DIGITAL-BASED MANAGEMENT IN THE HIGHER EDUCATION SYSTEM

Executives in all higher education sectors generally made forward-looking predictions about what would happen in the future and carried out strategic planning activities in this direction. However, in 2019, none of these executives predicted the global pandemic and its economic, social, and environmental impacts worldwide (van Rij 2015).

One way or another, there have been deviations in the strategic plans of all higher education institutions and a period has entered in which they had to cope with great difficulties. Most higher education institutions have been forced to make decisions that fundamentally change the way they do business and perhaps permanently transform business processes digitally. While the year 2020 is full of great uncertainty for everyone in general, priority issues arise in the field of cyber security (Deloitte 2018b).

The widespread use of remote work, which started with the pandemic, and the new normal business models, will require rapid digitalization. The pandemic has affected nearly all higher education institutions and the industry, forcing them to work better in significantly changing distance education and working conditions. Following this changing control environment, IT and IT security challenges will likely continue to increase in the future. In the longer term, it will cause high technology solutions to come into play. In this process, internal auditors should pay attention to returning to work, school, and other activities, potentially revealing new risks to privacy, "personally identifiable information (PII)" and "protected health information (PHI)" (Neale and Tryniecki 2020).

In response to these risks, the desire to reduce the risk of exposure may make it necessary to offer solutions that will accelerate the transition to autonomous vehicles and robotic solutions (EU, European Group on Ethics in Science and New Technologies 2018). Particularly, universities will give priority to the research and other project areas on this issue. In this context, "virtual private networks (VPN)" will continue to gain importance.

Higher education institution is one of the most basic institutions for the creation of human resources in a particular branch of science. Social life, economy, and technological developments lead to the emergence of new fields and new branches. In this respect, extensive development and knowledge accumulation arise in higher education. The HE institutions are the most important stage in the production of technology, especially for the application of knowledge from the past to life.

In this context, universities have been the means of scientific knowledge and transformation of this knowledge into technology. The development that was effective in the emergence of this process has been automation systems. The development of automation and the resulting production methods and techniques have accelerated with new advancements. Today, this effect shows itself together with digitalization (Schallmo et al. 2017).

However, digitalization is a multidimensional development that will reveal a profound structural change in relationships, communication, and institutional structures. The most important result of these developments is the social change they will bring about in institutions. While automation changes the subjects to be learned in higher education systems, digitalization has led to the intertwining of science branches and the emergence of interdisciplinary areas in an interaction (Fadeeva et al. 2014; Newman 2017; Schwertner 2017; McKinsey 2020). The common point of these interdisciplinary fields is computer and data analysis and the basic tools required for this. As a result, areas in which data-based knowledge is the basis of analysis are rapidly becoming integrated or initiating an interdisciplinary interaction through mathematics, statistics, and computer languages.

Data and data-based analysis in social sciences and scientific analysis based on data rapidly transform these fields into disciplines where data analysis is used as the basis. As a result, computer programs have become important in the use of mathematical statistics and most importantly, analysis tools. The widespread use of open-source analysis tools increases the need for this. For this reason, while these tools are included in the disciplines faster, this situation may cause young people or beginners to have more knowledge and to produce (CISCO 2016). While this process reveals the necessity of reviewing the duration of higher education, it also necessitates the consideration of the situation that may arise due to the closure of the knowledge gap between professors and learners (Allui and Sahni 2016; Ciesewicz and Ciesewicz 2018; Deloitte 2018a).

Universities that do not use funds from the state during the pandemic process face a significant financial problem (Smalley 2020). During this period, although students' main preferences are to study on campus, they accepted distance education because of this necessity. Considering the financing and cost of investments to be made by the HE institutions in areas such as distance education and digitalization, it can be expected that the liquidity problems will increase more in the next period (Hamzaee 2015; Kim 2020). The main reason for this is that it will take some time for economies to enter the growth process at the national and global levels because the new period requires new infrastructure investments to be made. If these new investments are not made, the sustainability of the HE institutions will not be possible in terms of both the competition process and the adaptation level of the development of technology (Terrisse 2020).

When evaluated from this point of view, as the use of financial resources and providing funds becomes of paramount importance during and after the pandemic process, the HE institutions will mainly switch to research projects and consultancy services. As a result, good planning, and development of relationships with external stakeholders will be required for the provision of teaching materials and new services or supports (Kettunen 2014).

Advisory fieldworks of internal audit departments will become the most important in making forward-looking strategic decisions in the planning of increasing financial resources and at the same time the required infrastructure (Bozkus Kahyaoglu and Coskun 2020). Especially in the pandemic period, planning for the use of resources and the control and auditing of the necessary records under the legal regulations in this period, together with the evaluations for the more effective use of resources in the future, will be the main source of information for management who will take decisions.

7.4 VIRTUALIZATION OF THE AUDIT UNIVERSE IN HIGHER EDUCATION

Developments regarding the digitalization of business processes and the emergence of a significant change in the working environment and interactions of internal audit departments have been mentioned in the previous chapters. In this context, risky and prioritized issues in university auditing arising from the changing audit universe and virtual environment are discussed in this chapter as follows.

In today's higher education sector, there is a trend toward internal audit functions to fulfill innovative, value-creating, and technology-supported tasks, beyond maintaining traditional assurance activities. With the transition to changing and digitalizing business models in higher education institutions, specifically because of the accelerated integration of new digital platforms in the pandemic, internal auditors are forced to develop new approaches to test digital risks and controls. Even if full digitalization is not present in this process, internal auditors are trying their best to activate remote auditing approaches by digital technologies (Beni 2019; Betti and Sarens 2020; Cangemi 2015).

This digital environment also makes it imperative for internal audit departments to turn to staff with the right digital skill mix (PwC 2019). In this context, since finding competent personnel in the sector is a laborious task, a new approach is also required in recruitment and training processes. In these circumstances, the audit committee must assist the chief audit executive (CAE) and the internal audit team to provide more value, especially in the framework of the oversight responsibility.

It should be in mind that corporate culture is a complex but exceptionally valuable "asset" by considering the scale and structure of the universities. When examining the effect of culture on the audit universe from the perspective of internal audit, although it is difficult to copy the corporate culture by looking at different institutions, it turns into a strong competitive advantage against other institutions when it is compatible with the corporate strategy (Nakiyaga and Anh 2017).

While revealing the risks in the audit universe, culture is considered as the first and primary issue that should be understood and internalized. Because, in order to provide realistic recommendations with added value within the framework of internal audit findings, it is necessary to take into account the corporate culture. While internal auditors evaluate the audit universe from the broadest perspective, they should closely monitor the way employees do business, how they interact with each other in the decisions they make, and how they achieve their goals.

Internal auditors must evaluate the corporate strategy and corporate culture impartially and independently and contribute to its transformation if needed. Particularly by understanding the current and desired future, and by activating various organizational levers such as changes in management processes, learning, recruiting, and leader behaviors, internal auditors could provide value-added views to direct the corporate culture toward the desired direction.

It is a fact that creating a corporate culture in the higher education sector is increasingly becoming a priority for Boards of Directors. In this context, internal audit, through its work and focus on audit results and cultural themes, can provide the Board with reasonable assurance and broad perspective on the state of the organizational culture and the sensitivity of the organization to accept and communicate cultural issues.

It is also critical to monitor attitudes toward targeted cultural initiatives to transform the organization into a desired organizational culture through the auditing and evaluation of corporate culture. With the contribution of an internal audit department with this perspective and responsibility, it may be possible for the corporate culture to realize an effective and efficient transformation process. This situation is more important than ever, especially in the pandemic period when digitalization is coming rapidly and inevitably entering the main business processes of the HE institutions.

The higher education sector is among the regulated markets and therefore the desire to gain competitive advantage, as well as regulatory pressures are influential in the decisions taken. In this respect, risk culture, as a component of corporate culture, is at the top of the agenda of many Boards of Directors. Accordingly, risk culture is increasingly seen as a priority measure for IA. In particular, "risk intelligence culture" is defined as the "invisible glue" (Deloitte 2018a, 2018b) that enables higher education institutions to work effectively and efficiently.

It is not enough for internal audit departments to focus on catching up with digital developments. However, adaptation is required as to how the audit will be carried out will also change. The fundamental change observed primarily in internal audit departments appears to be that they are starting to move into an advisory role based on continuous monitoring of controls, rather than providing assurance based on a retrospective point-to-time sample of what is already happening within the HE institution. With this new role, internal audit departments' perspective on the audit universe is also dramatically changing. IA now needs to address the tools, technologies, and methodologies of audits more comprehensively, and to double the digital direction before process owners.

In this respect, internal auditors must perceive the "context" in the HE institution very well. This means not only a good perception of the environment physically, but also the correct understanding and analysis of the data starting from the very first source in the digitalization process. Based on the data, internal auditors are required to use statistical modeling to build and test a model by making the existing situation meaningful and provide solutions for the virtual environment according to the findings obtained from this model.

For smart internal auditors who exploit smart systems, where the real action takes place, is to provide insights into the virtual audit universe and to detect emerging risks early. To achieve this, internal audit departments must focus on digitally augmenting the right capabilities with the right technology to manage risk and provide insights into their audit fieldwork. When focused right, it creates a reliable and sustainable audit model for the future. A high-level structure of what this might look like is summarized by Master and Bansal (2019) as a generic virtual audit universe model in Figure 7.4.

The difference of the virtual audit universe from the traditional auditing approach is shown in Figure 7.4; in the past, running this cycle required a human-based analysis where data was extracted, filtered, queried, analyzed, and reported each time. In the digitalized audit universe, internal auditors are supported by new, advanced tools such as advanced analytics, audit automation, machine learning, and artificial intelligence.

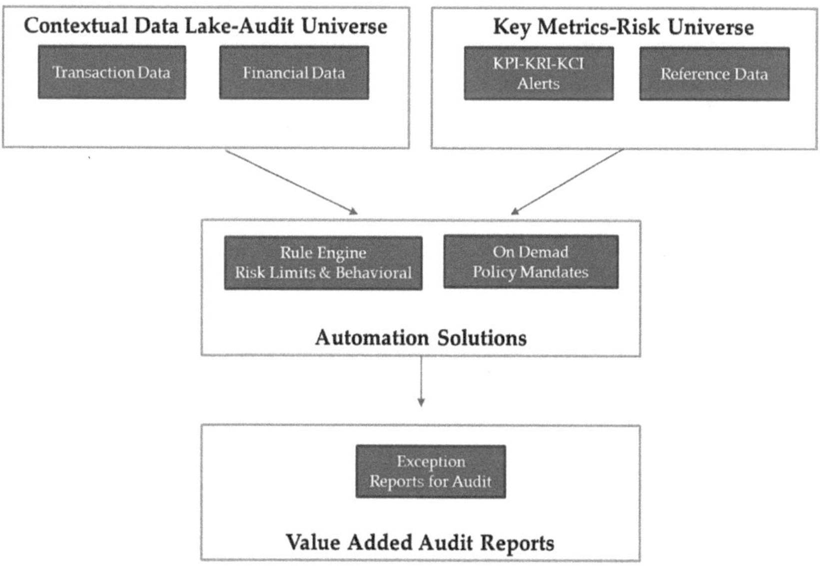

Figure 7.4 Virtual audit universe model for the future higher education.

Source: Adapted from Master and Bansal (2019).

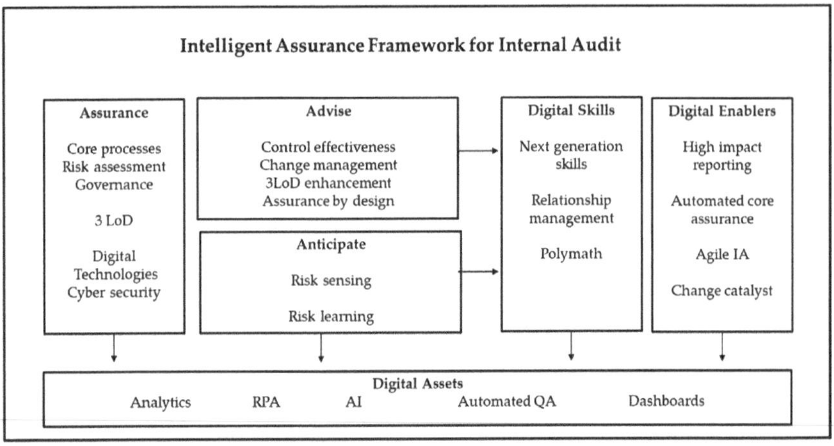

Figure 7.5 Triad of value approach for internal audit.

Source: Deloitte (2018a, 2018b).

The most important contribution of acquiring technology support is that now more time remains to focus on examining results that require human creativity, professional skepticism, and intelligence, because it is possible to analyze most of the heavy and monotonic work via technological tools and techniques under the guidance of the internal audit department.

With digitalization, it can be stated that the service scope and service delivery models of internal audit should be updated. However, the primary purpose and rationale of internal audit are almost the same; namely, to provide assurance and consult management to provide high value-added recommendations. However, to successfully achieve this main objective, internal audit functions are required to have predictions that will anticipate and help organizations adapt to and prevent emerging risks through proactive assurance. Deloitte defines this new approach as "internal audit 3.0" with three A's as "assurance", "advise", and "anticipate", i.e., a "triad of value". This is shown in Figure 7.5.

7.5 THE NEW SECURITY DIMENSION OF BEING VIRTUAL IN HIGHER EDUCATION

With the pandemic, a new era has emerged in which institutional structures for optimizing the balance between openness and cyber security in the world, gain importance. As a result of digitalization, an imbalance that may arise between the theft of information in the virtual environment and the application of openness in the institutional functioning creates the "cyber security" problem.

The higher education sector is one of the areas where openness and transparency policy is most important. In this respect, higher education institutions are faced with a significant financial and reputational risk with digitalization. It is a fact that all business processes and educational structures in HE institutions are based on digitalization makes it necessary to spend on the security as well as the infrastructure in order to maintain the operability of this system.

In addition, as the basic features of digital infrastructures are integrated with security, it may lead to higher expenditures in this area. The main characteristics of the higher education system and universities are the transparency of their information as well as the high level of information that should be protected.

It can be said that higher education institutions and research institutes, which prepare projects for industries and are stakeholders in important projects, face many cyber security problems. In this context, it is possible to evaluate cyber-attacks against them as "industrial espionage" (Deloitte 2015).

In addition, cybersecurity risks can be the most imperative consequence of reputation risk, as the factors that may arise from cyber-attacks that may affect the processes of students' assessment and evaluation processes will directly affect institutions as reputation risk. Therefore, it can be said that cybersecurity and related infrastructures are the main determining variables in maintaining the reliability of higher education institutions. In addition, it is important to continuously monitor cybersecurity risks via the internal audit department and to establish the necessary processes for taking precautions. It requires the creation of a system to prevent this process with an understanding and structure for full-time and instant monitoring, i.e., CA/CM approach (Bozkus Kahyaoglu and Coskun 2020; Valverde-Berrocoso et al. 2020). In this respect, internal audit departments, which are the main tool of risk management and have gained functionality with the new methods and techniques they apply, should be utilized.

The internal audit department can present critical information regarding cybersecurity to the management within the scope of its consultancy role. Thus, internal audit departments become a system for directly managing reputational risk. Therefore, internal audit departments should carry out internal audit activities to take necessary precautions against digital infrastructure and the deficiencies of the infrastructure, especially in the higher education system.

However, one of the major problems that may be encountered here will be the general adequacy of the internal auditors or internal audit department's knowledge in digital. The point to be considered here is that there is a very high rate of personal information in higher education institutions. For this reason, an organizational structure in which cyber security, risk management, and IT systems are effective is required in the higher education system. Continuous training is required to develop expertise in IT, digital and cyber risk management. The main feature of the personnel trained in this field is that the training process is quite long and demanding and requires continuous follow-up of new techniques.

It is a necessity for those working in this field to have basic training in computer skills, software, and hardware, as well as cybersecurity. The important point here is that there are few experts in cyber security, and it takes time to train personnel with the necessary skills, which will expose universities to a human capital problem in the future. In this respect, the high cost of consultancy services received in this field and high wages paid to the relevant staff will be an important source of risk for the HE institutions.

It is a fact that the offerings and qualifications of the HE institutions gain importance in e-learning. However, measuring and evaluating what has been learned, and most importantly, determining the criteria for the use of information in practice, and in this sense, measuring the competence of those who receive it comes to the fore as a problem. One of the new issues that the internal auditors should focus on will be the audit field work to determine the effectiveness of measurement and evaluation of the HE systems (Ciesewicz and Ciesewicz 2018; Fonseca et al. 2020). Although digital structures are a source from which much information can be obtained in general, it also raises the problem of using this information in a specific field, especially under "scientific ethics" (Origami Risk 2020).

Internal auditing should introduce new approaches and techniques to create safeguards against ethical risks that may arise as a risk management tool. In this respect, there is a reputational sensitivity against ethical risks and internal audit departments should be directed to a new audit engagement in this regard (Deloitte 2020).

Higher education institutions need to review their incident response "service level agreement (SLA)" competence, and internal audit departments should pay special attention to this issue. In addition, it is necessary to check that the processes for creating secure backups that they can return to when necessary are functional (Higher Education Authority – HEA 2017). With digitalization in the new era, organizations will become the target of more cyber-attacks. Therefore, it is very important that prevention and recovery strategies are in place and periodically audited.

NOTE

1 Please note that "Johns Hopkins Center for Health Security", "The Council for Higher Education Accreditation", and "Tuscany Strategy Consulting" have prepared a toolkit based on a joint work to assist the HE institutions for planning their future reopening strategies after COVID-19. Please visit https://www.opensmartedu.org/planning-tools/#guide.

REFERENCES

Alammary, A., Sheard, J., Carbone, A. (2014). Blended learning in higher education: Three different design approaches. *Australasian Journal of Educational Technology*, 30(4), 440–454. https://ajet.org.au/index.php/AJET/article/view/693/1061

Alexander, P. A. (2018). Information management versus knowledge building: Implications for learning and assessment in higher education. In O. Zlatkin-Troitschanskaia, M. Toepper, H. A. Pant, C. Lautenbach, C. Kuhn (Eds.), *Assessment of learning outcomes in higher education. Cross-national comparisons and perspectives* (pp. 43–56). Springer.

Allui, A., Sahni, J. (2016). Strategic human resource management in higher education institutions: Empirical evidence from Saudi. 12th International Strategic Management Conference, ISMC 2016, 28–30 October 2016, Antalya, Turkey. *Procedia – Social and Behavioral Sciences*, 235(2016), 361–371.

Alves A. de M., Harsh, M. (2018). *Project-and experience-based learning for communication skills development: Analysis of complementary studies courses for engineers*. In *Conference: 2018 3rd international conference of the portuguese society for engineering education (CISPEE)*. DOI: 10.1109/CISPEE.2018.8593464. https://www.researchgate.net/publication/330028580_Project-and_experience-based_learning_for_communication_skills_development_Analysis_of_complementary_studies_courses_for_engineers

Aristovnik, A., Damijana, K., Dejan, R., Nina, T., Lan, U. (2020). Impacts of the COVID-19 pandemic on life of higher education students: A global perspective. *Sustainability*, 12, 8438. DOI: 10.3390/su12208438. https://www.researchgate.net/publication/343555357_Impacts_of_the_COVID-19_Pandemic_on_Life_of_Higher_Education_Students_A_Global_Perspective

Beni, M. (2019). A simple revolution for digital auditing and auditing digital. https://www.icaew.com/technical/audit-and-assurance/faculty/audit-and-beyond/audit-and-beyond-2019/audit-and-beyond-december-2019/a-simple-revolution-for-digital-auditing-and-auditing-digital

Bentaa D., Bologaa, G., Dzitac, I. (2014). E-learning platforms in higher education case study. 2nd International Conference on Information Technology and Quantitative Management, ITQM 2014. *Procedia Computer Science*, 31, 1170–1176. E-learning Platforms in Higher Education. Case Study (sciencedirectassets.com)

Betti, N., Sarens, G. (2020). Understanding the internal audit function in a digitalised business environment. *Journal of Accounting & Organizational Change*, 17(2), 197–216. Emerald Publishing Limited 1832-5912. DOI: 10.1108/JAOC-11-2019-0114

Biggs, J. (1999). *Teaching for quality learning at university*. Buckingham: Open University Press.

Bozkus Kahyaoglu, S., Coskun, E. (2020). *Rethinking audit culture in digital transformation era: Recommendations for auditing digitally transforming universities*. In *The 11th International Conference on Governance Fraud Ethics and Corporate Social Responsibility (CSR)* which was virtually hosted by London Metropolitan University, Guildhall School of Business and Law on 27–28 August, 2020.

Bunar N. (2017). Migration and education in Sweden: Integration of migrants in the Swedish school education and higher education systems. https://nesetweb.eu/wp-content/uploads/2019/06/Migration-and-Education-in-Sweden.pdf

Caird, S., Roy, R. (2019) Sustainable higher education systems. In F. W. Leal (Ed.) *Encyclopedia of sustainability in higher education*. Cham: Springer. DOI: 10.1007/978-3-319-63951-2_261-1

Cangemi, M. P. (2015). *Staying a step ahead internal audit's use of technology*. The Institute of Internal Auditors Research Foundation (IIARF) Publications. https://www.researchgate.net/publication/281176816_Staying_a_Step_Ahead_Internal_Audit%27s_Use_of_Technology_by_Michael_P_Cangemi

Cardoso, A. R., Portela, M., Carla, S., Alexandre, F. (2007). Demand for higher education programs: The impact of the bologna process. CESifo Working Paper No. 2081. https://www.ifo.de/DocDL/cesifo1_wp2081.pdf

Carpentier, V. (2020). *COVID-19, long economic cycles and the prospects of a new World*. Global Center for Global Higher Education. https://www.researchcghe.org/blog/2020-04-30-covid-19-long-economic-cycles-and-the-prospects-of-a-new-world/

Cassidy, S. (2011). Self-regulated learning in higher education: Identifying key component processes. December 2011. *Studies in Higher Education* 36(8):989–1000. DOI: 10.1080/03075079.2010.503269. https://www.researchgate.net/publication/232926649_Self-regulated_learning_in_higher_education_Identifying_key_component_processes

Ciesewicz, E., Ciesewicz, J. (2018). The social capital approach. *College and University Auditor Journal*. Fall 2018. https://acua.org/College-and-University-Auditor-Journal/Fall-2018/The-Social-Capital-Approach

CISCO. (2016). The technology behind digital business strategy how leaders use data, connectivity, security, automation, and analytics to transform businesses. https://www.cisco.com/c/dam/m/en_us/never-better/assets/files/technology-behind-digital-business-strategy-tap.pdf

Davies, M., Devlin, M., Tight, M. (Eds.). (2010). *Interdisciplinary Higher Education: Perspectives and Practicalities*, Vol. 5. Bingley: Emerald. ISBN:978-0-85724-371-3. https://www.researchgate.net/publication/261951223_Interdisciplinary_Higher_Education_Perspectives_and_Practicalities

Deloitte. (2015). Cybersecurity: The changing role of audit committee and internal audit. https://www2.deloitte.com/content/dam/Deloitte/sg/Documents/risk/sea-risk-cyber-security-changing-role-in-audit-noexp.pdf

Deloitte. (2018a). Building on new approaches. 2019 Planning priorities for internal audit. https://www2.deloitte.com/content/dam/Deloitte/ie/Documents/Audit/IE_A_Building_on_new_approaches_A4_Portrait_TL_1118_FINAL_WEB.pdf

Deloitte. (2018b) Elevating cybersecurity on the higher education leadership agenda Increasing executive fluency and engagement in cyber risk. https://www2.deloitte.com/content/dam/Deloitte/us/Documents/public-sector/us-ps-elevating-cybersecurity-on-the-higher-education-leadership-agenda.pdf

Deloitte. (2020). Internal audit insights high-impact areas of focus – 2020. www2.deloitte.com/content/dam/Deloitte/ie/Documents/Risk/IE-risk-inernal-audit-insights-2020.pdf

DeVaney, J., Shimshon, G., Rascoff, M., Maggioncalda, J. (2020). Higher ed needs a long-term plan for virtual learning. *Harvard Business Review*. Summer, 1–6. https://hbr.org/2020/05/higher-ed-needs-a-long-term-plan-for-virtual-learning

Di Pietro, G., Biagi, F., Costa, P., Karpiński, Z., Mazza, J. (2020). The likely impact of COVID-19 on education: Reflections based on the existing literature and recent international datasets. EU Joint Research Center Technical Report.

ISBN:978-92-76-19937-3. https://publications.jrc.ec.europa.eu/repository/bitstream/JRC121071/jrc121071.pdf
Ehlers, U. D. (2020). Future skills – Future learning, future higher education. ISBN:978-3-658-29297-3, DOI: 10.1007/978-3-658-29297-3. https://nextskills.org/wp-content/uploads/2020/03/Future-Skills-The-Future-of-learning-and-higher-education.pdf
EU. (2020). The European Higher Education Area in 2020. Bologna Process Implementation Report. https://op.europa.eu/en/publication-detail/-/publication/c90aaf32-4fce-11eb-b59f-01aa75ed71a1/language-en/format-PDF/source-183354043
EU, European Group on Ethics in Science and New Technologies. (2018). Artificial intelligence, robotics and 'autonomous' systems. ISBN:978-92-79-80329-1. https://ec.europa.eu/research/ege/pdf/ege_ai_statement_2018.pdf
EUA. (2020). Academic freedom and institutional autonomy: Developments in Europe and beyond. https://eua.eu/news/598:academic-freedom-and-institutional-autonomy-developments-in-europe-and-beyond.html
European Foundation for Management Development (EFMD). (2020). In Nadia Elaref (Ed.), *Post pandemic paradigm of higher education*. Egypt: ESLSCA University. https://blog.efmdglobal.org/2020/07/07/post-pandemic-paradigm-of-higher-education/
Fadeeva, Z., Galkute, L., Mader, C., Scott, G. (2014). Assessment for transformation – Higher education thrives in redefining quality systems. In Zinaida Fadeeva, Laima Galkute, Clemens Mader, Geoff Scott (Eds.), *Sustainable development and quality assurance in higher education*. Palgrave Studies in Global Higher Education. ISBN:978-1-137-45914-5.
Filho, W. L., Raath, S., Lazzarini, B., Vargas, V. R., de Souza, L., Anholon, R., Quelhas, O. L. G., Haddad, R., Klavins, M., Orlovic, V. L. (2018). The role of transformation in learning and education for sustainability. *Journal of Cleaner Production*, 199, 286–295. DOI: 10.1016/j.jclepro.2018.07.017, ISSN 0959-6526. http://www.sciencedirect.com/science/article/pii/S095965261831984X
Fonseca, A.R., Susana, J., Caio, N. (2020). The role of internal auditing in promoting accountability in Higher Education Institutions. *Journal of Public Administration* 54(2), 243–265. https://www.scielo.br/pdf/rap/v54n2/1982-3134-rap-54-02-243.pdf
Gallagher, S., Palmer, J. (2020). The pandemic pushed universities online. The change was long overdue. *Harvard Business Review*. https://hbr.org/2020/09/the-pandemic-pushed-universities-online-the-change-was-long-overdue
Goos, M., Salomons, A. (2017). Measuring teaching quality in higher education: Assessing selection bias in course evaluations. *Research in Higher Education* 58, 341–364. DOI: 10.1007/s11162-016-9429-8. https://link.springer.com/article/10.1007/s11162-016-9429-8#citeas
Haapanen, M., Böckerman, P. (2013). Does higher education enhance migration? IZA Discussion Paper No. 7754. http://ftp.iza.org/dp7754.pdf
Hamzaee, R. (2015). An exploratory analysis of higher education financial challenges and innovations. *Journal of Business and Economics*, 6(4), 613–624. ISSN:2155-7950, USA. DOI: 10.15341/jbe(2155-7950)/04.06.2015/001. https://www.researchgate.net/publication/282468524_An_Exploratory_Analysis_of_Higher_Education_Financial_Challenges_and_Innovations

Hazelkorn, E. (2013). *Reflections on global problems of higher education: A European perspective*. NAFSA: Association of International Educators. https://www.nafsa.org/sites/default/files/ektron/files/underscore/reflections_on_global.pdf
HEA. (2017). Service level agreement Department of Education and Skills & Higher Education Authority. https://hea.ie/assets/uploads/2017/04/Service-Level-Agreement_2017_Department-of-Education-and-Skills__Higher-Education-Authority.pdf
Hevia, F. J., Vergara-Lope, S. (2019). Educational accountability or social accountability in education? Similarities, tensions, and differences. accountability working paper. https://accountabilityresearch.org/wp-content/uploads/2019/11/WP5-Educational-Accountability-Hevia-and-Vergara-Lope-ENG-web-14-Nov-2019.pdf
Institute for Teaching and Learning Innovation-(ITaLI). (2015). Future trends in teaching and learning in higher education. https://itali.uq.edu.au/files/1267/Discussion-paper-Future_trends_in_teaching_and_learning.pdf
Ion, G., Romiţă, I., Jorge, P.-V. (2014). Research-based teaching and learning in higher education: The perspective of postgraduate students. ICED 2014. http://www.iced2014.se/proceedings/1575_ION_IUCU_PALACIO.pdf
Jackson, N., Cooper, B. (2012). The Lifewide Learning, Education & Personal Developmente-book. http://www.lifewideebook.co.uk/uploads/1/0/8/4/10842717/introduction.pdf
Jin, Y., Chripa, S., Stephen R. (2015). The Role of higher education in promoting lifelong learning. ISBN:978-92-820-1194-2. https://unesdoc.unesco.org/ark:/48223/pf0000233592
Johnes, G., Johnes, J. (2016). Costs, efficiency and economies of scale and scope in the English higher education sector. *Oxford Review of Economic Policy* 32(4), 596–614.
Kane, G. C., Palmer, D., Nguyen Phillips, A., Kiron, D., Buckley, N. (2016). Aligning the organization for its digital future, *MIT Sloan Management Review*, 58(1), 1–27.
Kettunen, J. (2014). The stakeholder map in higher education. In Feng Tao (Ed.), *Society, Education and Psychology, International Proceedings of Economics Development and Research*, vol. 78. Singapore: IACSIT Press. DOI: 10.7763/IPEDR. http://www.ipedr.com/vol78/007-ICSEP2014-S00020.pdf
Kim, J. (2020). The integration of learning design and public health. Learning innovation. A space for conversation and debate about learning and technology. https://www.insidehighered.com/blogs/learning-innovation/integration-learning-design-and-public-health
Kiselova M. M., Oksana A. H., Svitlana, V. B., Oleksandra, O. T., Alla, I. C. (2020). International cooperation among tertiary educational institutions: trends and prospects. *International Journal of Higher Education*, 9(7), 356–366. ISSN 1927-6044 E-ISSN 1927-6052.
Koshal, R., Koshal, M. (1995). Quality and economies of scale in higher education. *Applied Economics*, 27(8), 773–778. DOI: 10.1080/00036849500000068
Leontyeva, I. A. (2018). Modern distance learning technologies in higher education: introduction problems. *EURASIA Journal of Mathematics, Science and Technology Education*, 14(10), 1–8. 1578 ISSN: 1305-8223 (online) Open Access Research Paper. https://doi.org/10.29333/ejmste/92284
Li, H. (2013). *The rising cost of higher education: A supply & demand analysis*. Leonard N. Stern School of Business New York University. https://www.stern.nyu.edu/sites/default/files/assets/documents/con_042986.pdf

Marton, F., Saljo, R. (1997). Approaches to learning. In F. Marton, D. Hounsell & N.J. Entwistle (Eds.), *The experience of learning*. Edinburgh: Scottish Academic Press.

Master and Bansal. (2019). Impact of digital on the future of internal audit. EXL White Paper. https://www.exlservice.com/resources/assets/library/documents/EXL-WP-Impact-of-Digital-on-the-Future-of-Internal-Audit.pdf

McKinsey. (2020). COVID-19 and US higher education enrollment: Preparing leaders for fall. https://www.mckinsey.com/industries/public-and-social-sector/our-insights/covid-19-and-us-higher-education-enrollment-preparing-leaders-for-fall

Nakiyaga, B., Anh, D. T. L. (2017). How organizational culture affects internal control effectiveness: The role played by top management case study: Uganda Revenue Authority. Business Administration Master's Thesis Karlstad Business School. https://www.divaportal.org/smash/get/diva2:1114871/FULLTEXT01.pdf

National Association of Independent Colleges and Universities-NAICU. (2020). Safe on campus: A framework for reopening colleges and universities. https://www.wpi.edu/sites/default/files/2020/05/27/Higher%20Ed%20Framework%20Briefing%20for%20Posting.pdf

Neale, M., Tryniecki, M. (2020). The post-pandemic evolution of student data privacy. *EDUCAUSE*. https://er.educause.edu/-/media/files/articles/2020/8/er20_3104.pdf

Newman, C. (2017). Data analytics is driving efficiency in higher ed. *The College and University Auditor Journal* 1–6. https://acua.org/College-and-University-Auditor-Journal/Fall-2017/Data-Analytics-is-Driving-Efficiency-in-Higher-Ed

Nortvig, A. M. (2014). The change of time and space in E-learning. *American Journal of Educational Research*, 2(8), 612–616. http://pubs.sciepub.com/education/2/8/9/index.html#

Nusche, D. (2008). Assessment of learning outcomes in higher education: A comparative review of selected practices. OECD Education Working Paper No. 15. http://www.oecd.org/australia/40256023.pdf

OECD. (2008). Higher education to 2030. The reversal of gender inequalities in higher education: An on-going trend. ISBN:978-92-64-04065-6. Volume 1: Demography. http://www.oecd.org/education/ceri/41939699.pdf

OECD. (2014). Promoting skills for innovation in higher education: A literature review on the effectiveness of problem-based learning and of teaching behaviors. OECD Education Working Paper No. 100. Prepared by Sabine Hoidn and Kiira Kärkkäinen. https://www.oecd.org/education/ceri/Promoting-Skills-for-Innovation-in-Higher-Education.pdf

OECD. (2020). Learning environment. https://gpseducation.oecd.org/revieweducationpolicies/#!node=41710&filter=all

OECD IMHE. (2012). Fostering quality teaching in higher education: Policies and practices. https://www.oecd.org/education/imhe/QT%20policies%20and%20practices.pdf

Origami Risk. (2020). Facing the challenge of reputation management in higher education. https://www.origamirisk.com/sites/default/files/file/2020-09/Challenge-of-Reputation-Management-Higher-Ed-EB.pdf

Otola, I., Grabowska, M. (2020). Business models: Innovation, digital transformation, and analytics. In Iwona Otola, Marlena Grabowska (Eds.), *Data analytics applications*. CRC Press. ISBN:1000097773, 9781000097771.

Pianko, D., Jarrett, J. (2012). *Early days of a growing trend: Nonprofit/for-profit academic partnerships in higher education*. Boca Raton: Game Changers: Education and Information Technologies. ISBN:978-1-933046-00-6. Published by EDUCAUSE, https://www.educause.edu/-/media/files/library/2012/5/pub72037-pdf.pdf?la=en&hash=E9CC02315071C806FA6A4C44478DC196E3CE0CDE

Pramanik. (2014). Role of interdisciplinary studies in higher education in India. *Journal of Education and Human Development* 3(2), 589–595. ISSN: 2334-296X (Print), 2334-2978 (Online). http://jehdnet.com/journals/jehd/Vol_3_No_2_June_2014/34.pdf

PwC. (2019). Elevating internal audit's role: The digitally fit function 2019 State of the Internal Audit Profession Study. https://www.pwc.ru/en/riskassurance/assets/2019-state-of-the-internal-audit-profession-study-en.pdf

Rapaccioli, D. (2019). The role of higher education in today's global economy. By Donna Rapaccioli, Dean and University Professor at Gabelli Business School, Fordham University. https://newestcorp.com/higher-education-global-economy/

Ribeiro, F. A., Relvas, J. B. (2018). Profiling interdisciplinarity in higher education: an ecological approach. *Working Papers in Higher Education Studies*, 3, 121–145. https://www.wphes-journal.eu/index.php/wphes

Rogoff, K. (2020). An economist explains how COVID-19 will impact universities. *World Economic Forum*. https://www.weforum.org/agenda/2020/07/will-universities-learn-from-lockdowns

Rumbley, L. E., Altbach, P. (2016). The local and the global in higher education internationalization. In E. Jones, R. Coelen, J. Beelen, H. Wit (Eds.), *Global and Local Internationalization* (pp. 7–13). Transgressions: Cultural Studies and Education. Rotterdam: SensePublishers. DOI: 10.1007/978-94-6300-301-8_2. https://www.researchgate.net/publication/301265932_The_Local_and_the_Global_in_Higher_Education_Internationalization

Ruso, J., Horvat, A., Đurić, M., Trajković, A. (2015). In *The international conference on education and modern educational technologies (EMET 2015)*, 2015 July 16–20 (pp. 135–139), Zakynthos Island, Greece. ISSN:2227-4618, ISBN:978-1-61804-322-1.

Schallmo, D., Williams, C. A., Boardman, L. (2017). Digital transformation of business models—Best practice, enablers, and roadmap. *International Journal of Innovation Management*, 21(08), p. 1740014.

Schwertner, K. (2017). Digital transformation of business. *Trakia Journal of Sciences*, 15(1), pp. 388–393.

Sivagurunathan, S. (2012). An urgent global need of introducing multidisciplinary approaches in the humanities at the universities. *International Journal of Multidisciplinary Research*, 2(7), 14–28. ISSN:2231 5780. http://zenithresearch.org.in/images/stories/pdf/2012/JULY/ZIJMR/2_ZIJMR_JULY12_VOL2_ISSUE7.pdf

Smalley, A. (2020). *Higher Education Responses to Coronavirus (COVID-19)*. NCSL. https://www.ncsl.org/research/education/higher-education-responses-to-coronavirus-covid-19.aspx

Tan, A. H. T., Birgit, M., Anita, Z. (2016). A systematic review of quality of student experience in higher education. *International Journal of Quality and Service Sciences*, 8(2), 209–228.

Teijeiro-Álvarez, M., Rodríguez-Vázquez, C., Blázquez-Lozano, F. (2016). The importance of social capital in higher education. A study of the Facebook Fan pages. In Á. Rocha, A. Correia, H. Adeli, L. Reis, & M. Mendonça Teixeira (Eds.),

New advances in information systems and technologies. Advances in intelligent systems and computing, vol. 445. Cham: Springer. https://doi.org/10.1007/978-3-319-31307-8_48

Terrisse, A. (2020). How has the pandemic changed the face of edtech? https://www.eu-startups.com/2020/09/how-has-the-pandemic-changed-the-face-of-edtech/

Tomlinson M. (2015). Between instrumental and developmental learning: Ambivalence in student values and identity positions in marketized UK higher Education. *International Journal of Lifelong Education*, 34(5), 569–588. https://doi.org/10.1080/02601370.2015.1077482

Tomlinson, M. (2016) The impact of market-driven higher education on student-university relations: Investing, consuming and competing. *Higher Education Policy*, 29, 149–166. DOI: 10.1057/hep.2015.17

Toom, A., Kynäslahti, H., Krokfors, L., Jyrhämä, R., Byman, R., Stenberg, K., et al. (2010). Experiences of a research-based approach to teacher education: Suggestions for future policies. *European Journal of Education*, 45(2), 331–344.

Toutkoushian, R. K. (2016). *Revisiting economies of scale in higher education. The Meeting of the Association for Education Finance and Policy (AEFP)*, Denver, CO, March 17–19, 2016. https://aefpweb.org/sites/default/files/webform/41/Revisiting%20Economies%20of%20Scale%20and%20Scope%20in%20Higher%20Education.pdf

UNESCO IESALC. (2019). *Mobility in higher education in Latin America and the Caribbean: Challenges and opportunities of a renewed Agreement for the recognition of studies, titles and diplomas*, vol. 1. Caracas: I UNESCO International Institute for Higher Education in Latin America and the Caribbean (IESALC).

UNESCO International Institute for Educational Planning-IIEP. (2020). SDG 4 – Policies for flexible learning pathways in higher educationtaking stock of good practices internationally. Working paper. https://unesdoc.unesco.org/ark:/48223/pf0000372817

UNESCO International Institute for Higher Education (IESALC). (2020). COVID-19 and higher education: Today and tomorrow. Impact analysis, policy responses and recommendations. https://www.iesalc.unesco.org/en/wp-content/uploads/2020/05/COVID-19-EN-130520.pdf

University Industry Innovation Network (UIIN). (2020). Building the university of tomorrow to bridge higher education divide. https://uiin.org/2020/07/23/building-the-university-of-tomorrow-to-bridge-higher-education-divide/

Valverde-Berrocoso, J., María del Carmen, G.-A., Carmen, B.-V., María Belén, M.-C. (2020). Trends in educational research about e-Learning: A systematic literature review (2009–2018). *Sustainability*, 12, 5153. DOI: 10.3390/su12125153. https://www.mdpi.com/journal/sustainability

Van Rij (2015). 21st Century Higher Education: Quick scan of foresight and forward looks on higher education in the ICT age. UNESCO discussion paper. https://iite.unesco.org/files/news/639201/Foresight_on_HE_and_ICT_Discussion_paper.pdf

Vasarhelyi, M. A., Cho, S., Cheong, A., Chanyuan (Abigail), Z. (2020). Smart audit: The digital transformation of audit. *Journal of EU Court of Auditors*, 1, 27–33. Big Data & Digital Audit. No. 1. 2020.

Watson, W., Watson, S. (2014). Redesigning higher education: Embracing a new paradigm. *Educational Technology*, 54(3), 47–51. Retrieved January 11, 2021, from http://www.jstor.org/stable/44430273

Concluding remarks

The strategic role of higher education institutions is similar wherever they are in the world. In this context, the quality of their education and research work processes to increase social benefit is decisive. When the functionality of the audit processes of universities is ensured, reasonable assurance will be provided for the execution of transactions under the standards of internal control, risk management, and segregation of duties. The important point here is to ensure the competence of those who carry out the audit and the processes in which the audit is carried out. In this respect, the topics covered in this book reveal the change in audited processes due to digital transformation. Thus, this book is intended to contribute to the determination and presentation of the auditor profile that can keep up with the said change and the auditor's needs for this.

Among the topics covered in this work are the development and digital transformation process of universities over time from the auditing perspective, and the effects of technological developments are also closely examined. The main reason for this is to question the main factors, needs, and best practices of the developments in the audit universe and the risk universe, where the auditors carry out field works. Information on the competency areas required for the auditors to meet the necessary qualification and work area standards is presented. In addition, the processes of universities as an institution and the structure of their sub-processes are explained within the framework of the e-business model, and recommendations are made on how they should be audited. In this respect, new trends and future expectations about the audit profession are explained based on CA/CM approaches. In this study, the importance of cooperation between the auditee and the auditors is emphasized. Thus, directing to strategic goals, monitoring, and measuring the level of performance realization will be more effective. The university administration, that tries to detect the risks early and make the necessary revisions, should produce practical solutions based on the impartial, independent, and objective audit findings and recommendations they provide. We hope that this work, which is prepared to incorporate information that can be utilized by all corporate managers, executives, and supervisors and who aim to rank as an innovative, entrepreneurial, and university researcher in a globally competitive environment, will contribute to the entire scientific world.

DOI: 10.1201/9781003093008-8

APPENDIXES

Appendix 1 Self-Assessment for Maturity of Digital Internal Audit

Maturity	*Exploring*	*Building*	*Innovating*	*Disrupting/champions league*
Strategy	A small number or pilot projects have been performed, possibly with some success.	Future state definition of data-driven IA is clear and supported with a roadmap and ambition level.	The role of data is fully defined. Data are fully utilized for Business and IT audits driving strong outcomes supported with a strong culture & transformed IA.	Risk assessment and audit execution are largely driven by technology and data techniques supporting auditor judgment.
People	A small number or pilot projects have been performed.	IA Leadership drives digital, automation, and data analytics as a key initiative.	6–10% of FTEs with a mix of technical and non-technical backgrounds; quantitative/ data science skills on the team. Heavy commitment to training.	Risk assessment and audit execution are largely driven by technology and data techniques supporting auditor judgment. Training is tied to performance goals.
Process	Little to no cooperation across the enterprise. IA methodology does not leverage automation or mention data analytics.	The IA methodology specifically notes integration points and procedures related to data analytics integration.	IT supports improved infrastructure and access to data. Strong workflow and PMO procedures to support the data analytics team.	The IA data analytics team employs leading practices from various disciplines for automation, AI, RPA. Fully integrated data analytics activity workflows.
Data	Limited or no use of data within audits.	Basic data visualization capabilities are available.	Data from different sources is combined to perform advanced analytics.	All organizational data are being used to provide insights and assurance.
Technology	IA tools for collaboration, audit execution, and reporting are not in use. Data joins and filters support rules-based testing in Excel spreadsheet format.	The majority of work performed is script-based; basic routines may exist to obtain and manipulate data as a part of the scripts.	Advanced visualization, advanced analytics techniques, including predictive models, network analysis, and unsupervised techniques.	Cognitive capabilities (AI) and RPA are fully applied. Auditors rely heavily on results to drive planning and scoping activities. Robust procedures expedite access to data.

Source: Deloitte, 2020.

Appendix 2 Self-Assessment for Maturity of Strategy for Internal Audit

			Limited	Developing	Defined	Advanced	Leading	
Analytics vision	A documented and published insights strategy that is adopted by the business and IT communities and sets a vision and roadmap for exploiting the organization's data assets.	No insight strategy or governance framework exists.						There is a mature, documented, and published insights strategy that supports the analytic and business community and is sponsored by the leadership team.
Value & business case	How well are benefits measured and managed to ensure analytics meets financial targets, expectations and continues to deliver value?	Benefits have not been identified and documented.						Benefits are clearly identified, recognized across the organization, and have business owners. Tracking is an ongoing process with clear visibility and involvement. Analytics acts as a service to help achieve business-driven initiatives and is effectively self-funded in this context.
Stakeholder management	Key stakeholders are aligned on the insights vision and in support of the journey to becoming insight-driven.	Little knowledge of analytics vision outside of the primary stakeholder.						Stakeholders are actively engaged in and measured against the success of the insights agenda or vision.

(*Continued*)

Appendix 2 (*Continued*) Self-Assessment for Maturity of Strategy for Internal Audit

			Limited	Developing	Defined	Advanced	Leading	
Operating model	An analytics target-operating model has been tailored to the needs of the organization and is supported by structured governance and processes.	BI and analytics are conducted entirely ad-hoc with little alignment to a common objective or visible defined structure.						The operating model effectively uses capabilities throughout the organization, is scalable and agile, and is supported by well-understood roles, responsibilities, and controls.
Innovation	A culture of continuous innovation is embedded into the organization, enabling it to adapt to market changes.	Organization struggles to keep up with market disruptions						Organization rigorously challenges status quo and has established feedback loop for transformation programs.
Leadership	There is a champion for analytics at the executive level who can drive organizational change.	The analytics agenda does not have an identified champion at the executive level.						An analytics champion and their team consistently mobilize the organization around insights and innovation.

(*Continued*)

Appendix 2 (*Continued*) Self-Assessment for Maturity of Strategy for Internal Audit

			Limited	Developing	Defined	Advanced	Leading	
Organization design	The degree to which the roles, responsibilities, performance management, and organizational hierarchy support the collection, dissemination, and use of insights within the overall enterprise.	Organizational structure provides no support for insights.						The organization design fully supports the chosen operating model and provides attractive career paths for employees engaging with insights.
Talent	The organization has the right mix of technical, analysis, communication, and business acumen to deliver the end-to-end insight process.	Little or no insight or analytics training and skills exist within the organization.						Integrated specialist teams are regularly deployed throughout the organization to deliver insights. Analytics awareness is high and insights training is incorporated into learning pathways and specialist insights skills are embedded and rewarded in the Business.
Change journey	A change management journey has been planned to support the transformation into an insight-driven organization and data-driven decision-making forms a key part of the organization's culture	Not enough consideration has been given to change management in the context of analytics.						A top-down and bottom-up change management program has been implemented and the culture transformed to that of an IDO.

(*Continued*)

Appendix 2 (*Continued*) Self-Assessment for Maturity of Strategy for Internal Audit

			Limited	Developing	Defined	Advanced	Leading	
Knowledge management	There is a central knowledge-management repository that is used to support insights resources across the organization.	No knowledge-management capability exists. Knowledge maintained on personal hard drives with no version control or community portal.						Central social platforms exist and are widely used across the organization and continuous improvement in the insights process.

Source: Deloitte, 2020.

Appendix 3 Self-Assessment for Maturity of People for Internal Audit

			Limited	Developing	Defined	Advanced	Leading	
Leadership	There is a champion for analytics at the executive level who can drive organizational change.	The analytics agenda does not have an identified champion at the executive level.						An analytics champion and their team consistently mobilize the organization around insights and innovation.
Organization design	The degree to which the roles, responsibilities, performance management, and organizational hierarchy support the collection, dissemination, and use of insights within the overall enterprise.	Organizational structure provides no support for insights.						The organization design fully supports the chosen operating model and provides attractive career paths for employees engaging with insights.
Talent	The organization has the right mix of technical, analysis, communication, and business acumen to deliver the end-to-end insight process.	Little or no insights or analytics training and skills exist within the organization.						Integrated specialist teams are regularly deployed throughout the organization to deliver insights. Analytics awareness is high and insights training is incorporated into learning pathways and specialist insights skills are embedded and rewarded in the business.

(*Continued*)

Appendix 3 (*Continued*) Self-Assessment for Maturity of People for Internal Audit

		Limited	Developing	Defined	Advanced	Leading
Change journey	A change management journey has been planned to support the transformation into an insight-driven organization and data-driven decision-making forms a key part of the organization's culture.	Not enough consideration has been given to change management in the context of analytics.				A top-down and bottom-up change management program has been implemented and the culture transformed to that of an IDO.
Knowledge management	There is a central knowledge-management repository that is used to support insights resources across the organization.	No knowledge-management capability exists. Knowledge maintained on personal hard drives with no version control or community portal.				Central social platforms exist and are widely used across the organization and continuous improvement in the insights process.

Source: Deloitte, 2020.

Appendix 4 Self-Assessment for Maturity of Process for Internal Audit

			Limited	Developing	Defined	Advanced	Leading	
Ideation & prioritization	Business users are regularly consulted on ways to use analytics to solve issues and their needs are prioritized accordingly.	There is little to NO interaction with business users.						Business and analytics teams meet regularly to generate ideas for valuable analytics projects and rigorously score and test them.
Agility and scalability	The analytics-operating model is designed to handle the increased demand for services.	Analytics operating model and controls only take into account existing demand.						Analytics is delivered "as a service", able to respond and scale to demand flexibly from both business and IT perspectives, taking advantage of multiple delivery models, while keeping costs at a manageable level.
Process re-engineering	End-to-end enterprise processes evolve and mature through the use of insights gained from specific projects.	Analytics solutions are conducted in isolation and are not embedded into business as usual.						Analytics teams provide advisory services to business units to change and mature their existing business processes.

(*Continued*)

Appendix 4 (Continued) Self-Assessment for Maturity of Process for Internal Audit

			Limited	Developing	Defined	Advanced	Leading	
Governance	The processes in place to initiate, manage, maintain, and exploit Analytics as an enterprise resource, for example, work intake process, project management, project tracking, prioritization processes, etc.	No governance exists that defines the intake process for new Analytics requests, prioritization of activities or allocates time of analyst community, etc.						An Analytics governance process exists that manages the intake process, evaluates requests and selects, and prioritizes projects. The process is well understood and communicated and group leadership plays an active role in soliciting and suggesting projects across the organization.
Benefits Realization	The degree to which key goals and objectives of the organization have been defined in a well-communicated set of KPIs that can be tracked, measured, evaluated, and refined.	Key goals and objectives of the organization and the insights that it develops are not well defined or communicated.						Key goals and objectives of both the organization and the insights it develops are well defined and communicated in a set of KPIs that can be tracked, measured, evaluated, and refined.

Source: Deloitte, 2020.

Appendix 5 Self-Assessment for Maturity of Data for Internal Audit

		Limited	Developing	Defined	Advanced	Leading
Inf. model & sources	An information model which takes advantage of internal and external data sources that are easily accessible and structured in a way which will support the creation of trustworthy insights.	Little is known about the information and data held within the organization.				The information model is structured to allow for significant variation in data volumes and varieties and the data are accessible on-demand.
Data Quality	The degree to which the organization can trust the accuracy of BI data and proactively drives toward ensuring data are fit for purpose.	Risk information/ data quality unknown. No formal initiative to identify, cleanse data, and prioritize known issues. Data quality and control needs are addressed on an ad-hoc basis and based on pressing needs.				Risk information/data quality is trusted by the business. Unstructured information (e.g. documents) is also subject to quality controls. Data are tagged with quality indicators to identify known problems. Data quality metrics and rules are in place to ensure data are fit for purpose as defined by business data users.

(*Continued*)

Appendix 5 (*Continued*) Self-Assessment for Maturity of Data for Internal Audit

			Limited	Developing	Defined	Advanced	Leading	
Data Monetization	Data are recognized as a valuable asset for the organization and a clear policy exists as to how it will be used to generate revenue or savings.	No discussion has taken place around the value of data or whether it will be bought or sold.						A clear view has been defined on the role that data monetization plays both now and in the future.
Ethics & Sharing	The policies surrounding the way that customer and employee data will be used have been formalized, shared, and accepted by individuals.	No policy is in place which considers the ethical use of data.						The organization is known both internally and externally for exceptional ethical treatment and use of data, and is trusted by customers and employees alike.
Regulation & compliance	There is a focus on securing enterprise data assets from any unauthorized infringement to ensure that appropriate data security and access policies, checks, and controls are monitored.	There are no information privacy or security policies. Roles and responsibilities are determined largely in an ad-hoc fashion.						Enterprise-wide privacy and security policies are established and enforced. Policies are periodically reviewed and updated in order to stay ahead of regulators.

Source: Deloitte, 2020.

Appendix 6 Self-Assessment for Maturity of Technology for Internal Audit

			Limited	Developing	Defined	Advanced	Leading	
Reference architecture	The level of sophistication of both the physical hardware and software needed to support analytics development and the ease with which tools are quickly and effectively adaptable.	The analytics production environment is limited or nonexistent.						The analytics production environment includes all of the following tools or capabilities: advanced data storage, real-time process, ETL, visualization, etc., and includes a development timeline that upgrades systems to keep pace with the user community and business needs.
Tech disruptors & vendor strategy	Incorporation of new disruptive technologies is built into the IT strategy and sophisticated vendor management exists.	Organization lags behind in the adoption of new technology and does not act as an intelligent client.						An integrated technology ecosystem has been developed where the organization acts as an intelligent client, managing costs while remaining innovative.

(*Continued*)

Appendix 6 (*Continued*) Self-Assessment for Maturity of Technology for Internal Audit

			Limited	Developing	Defined	Advanced	Leading	
Discovery zone	Analytics testing and solutioning are conducted outside the production environment in a specially designed sandbox.	There is no sandbox environment or testing of analytics tools and solutions.						The analytics sandbox environment is set up specifically to deliver priority analytics solutions while testing disruptive technology and guides the future IT operating model.
Cloud vs. on-premise	The use of cloud-based applications or processing power for performing business analytics. Data integration, data modeling, query and reporting, dashboards, and advanced analytics.	The organization does not use Cloud services for Business Analytics.						The cloud is used for cleansing and improving data quality, creating data models, and analyzing data for critical business functions. This includes in-memory to detect correlations and patterns in very large datasets in seconds instead of weeks.

(*Continued*)

Appendix 6 (*Continued*) Self-Assessment for Maturity of Technology for Internal Audit

		Limited	Developing	Defined	Advanced	Leading
Security, reliability, & continuity	Analytics IT systems are stable and can support additional capacity without affecting performance.	IT systems experience latency and performance is consistently poor. Any broadening of the user base compounds the problem.				IT systems are available at all times required by the business, and can respond to increased loads in peak times without performance degradation or substantial costs.

Source: Deloitte, 2020.

Index

Page numbers in **bold** indicate tables.

G

H

I

T

U

V

W

X